SOLVING THE RIDDLE OF
PHYLLOTAXIS

Why the Fibonacci Numbers and the
Golden Ratio Occur on Plants

T0349876

SOLVING THE RIDDLE OF PHYLLOTAXIS

Why the Fibonacci Numbers and the Golden Ratio Occur on Plants

IRVING ADLER

Foreword by
Stephen L. Adler
Institute for Advanced Study, USA

Diagrams by
Peggy Adler

 **World Scientific**

NEW JERSEY · LONDON · SINGAPORE · BEIJING · SHANGHAI · HONG KONG · TAIPEI · CHENNAI

Published by

World Scientific Publishing Co. Pte. Ltd.

5 Toh Tuck Link, Singapore 596224

USA office: 27 Warren Street, Suite 401-402, Hackensack, NJ 07601

UK office: 57 Shelton Street, Covent Garden, London WC2H 9HE

British Library Cataloguing-in-Publication Data
A catalogue record for this book is available from the British Library.

Cover Illustration:

The graphic of the Golden Ratio was adapted by Peggy Adler, from a photograph she took of the custom crafted pendant that the author, Irving Adler, had made for him by a Bennington, Vermont jeweler.

ISBN 978-981-4407-62-5

Printed in Singapore by Mainland Press Pte Ltd.

For Roo,
who would have been
proud of this
family collaboration.

Foreword

The papers in this volume represent the third stage of my father's remarkable career, first as a teacher of mathematics, then as an author of outstanding science and mathematics books for children and adults, and finally as a mathematician and scientist studying phyllotaxis.

One of my vivid childhood recollections is the mathematics assembly program that my father wrote, incorporating an excerpt from Gilbert and Sullivan's "Mikado". My father at that time was Chairman of Mathematics at Textile High School in Manhattan. The program began with the Mikado singing his famous refrain, "My object all sublime, I shall achieve in time, to let the punishment fit the crime...", and then went on to the verse in which the Mikado promises that the "billiard sharp" shall be forced to play on "a cloth untrue, with twisted cue, and elliptical billiard balls!". After a brief dialog by the skit's actors, in the course of which an elliptical billiard ball — an egg — was shattered, the skit's announcer informed the audience that in the next scene, "The Billiard-Player's Dream", "All we have to do is change the Mikado's idea a little bit. Instead of using an elliptical billiard ball, we'll use an elliptical billiard table".

My father had asked the school's shop teacher to make an elliptical billiard table, which the audience could see by means of an angled mirror suspended over it. The cast in the skit demonstrated that a billiard ball placed at one focus of the ellipse, when shot towards the perimeter, would bounce off and always hit a billiard ball placed at the other focus. The skit then went on to deal with other properties of conic sections. This assembly program was needless to say a hit with the students, and was repeated a number of times. I think this program encapsulates my father's approach to education, which was to motivate students to *want* to come to grips with the really interesting content of the subject.

In my own career, I benefited greatly from my father's approach to teaching. When I was quite young, my father built me electrical toys such as as telegraphs, a "burglar alarm" that rang a bell when a door was opened, and a miniature traffic light. Later on, when I became interested in radio while in elementary school, my father gave me a copy of Marcus and Marcus's classic World War II text "Elements of Radio"; my father made a point of letting me be the family radio expert, while he was the consultant for the few bits of algebra in the text. Also at my father's suggestion, I started to canvass the neighborhood door to door, pulling a small wagon and asking for old radios, appliances, and television sets people were planning to discard. I stripped the parts out of these, and used them to build radios, amplifiers, and even an oscilloscope using a salvaged 7-inch television tube. Much later, when I had just finished high school, I decided to teach myself calculus in spare time from my job at Bell Laboratories in Manhattan, during the summer before college. My father gave me his old calculus text, along with the sage advice to do every *third* problem — because I had to do problems to learn the material, but there was not time (and it would be too boring) to try to do all of them. In all of these things, my father's pedagogical approach was to engage my curiosity and to empower me to do things on my own, just as he had with his high school students.

My father's transition from teacher to author was stimulated by events from the outside: during the McCarthy era witchhunts, he was dismissed from his job in the New York City schools. Within two years he had reestablished himself in a new, and what turned out to be a very successful career, as writer of popular mathematics and science books for children and adults, many published by the John Day Company and by Knopf. His first John Day book, "Time in Your Life", was illustrated by my mother Ruth, and she continued to do illustrations for subsequent books. My sister Peggy, who compiled this volume and illustrated my father's papers, also illustrated a number of the books. After my parents moved to Bennington, Vermont in 1960, my mother started a series of books for younger children, the "Reason Why" series, that she coauthored with my father. Altogether, my father wrote 56 books on mathematics and science, and another thirty coauthored with my mother. Many of the books are classics, and I expect that in the future a number of them will be reprinted.

While I was at college in the late 1950s my father decided to return to Columbia, from which he had received a Master's degree in mathematics in 1938, to pursue a doctorate in mathematics. He received his PhD in 1961, writing a thesis on composition algebras under the direction of Ellis Kolchin. However, he did not continue to work in this abstract area of mathematics. Instead, after moving to Bennington, he became

interested in the classic problem of phyllotaxis, the spiral arrangement
of leaves on plants, and its intimate and perplexing connections with
such number theoretic and geometric concepts as the Fibonacci series
of numbers, and the golden section. His research on phyllotaxis led to
papers published in the *Journal of Theoretical Biology* and elsewhere,
that are major and enduring contributions to the field. These papers
are collected in this volume, and are a late, and important facet, of
my father's muli-faceted career. They are a tribute to the intellectual
curiosity, mathematical and scientific skill, and sense of purpose
that underlay his many years as a teacher, an author, and a scientific
researcher.

Stephen L. Adler
Princeton, New Jersey
March 3, 2012

Contents

Contents

Credits

1) The Preface by Stephen L. Adler, Professor Emeritus at the Institute for Advanced Study, Princeton, is drawn in part from his essay, "From Elements Radio to Elementary Particle Physics" in the volume, *One Hundred Reasons to be a Scientist*, published in 2004 by the Abdus Salam International Center for Theoretical Physics and in part from the remarks he made on the occasion of the first Ruth and Irving Adler Expository Lecture in the School of Mathematics at the Institute for Advanced Study, February 2000.

2) "A Model of Contact Pressure in Phyllotaxis" is a review article, originally published in the *Journal of Theoretical Biology*, Volume 45, Issue 1, May 1974, pp. 1–79, © Elsevier.

3) "A Model of Space Filling in Phyllotaxis" is an original research article published in the *Journal of Theoretical Biology*, Volume 53, Issue 2, 1975, pp. 435–444, © Elsevier.

4) "The Consequences of Contact Pressure in Phyllotaxis" is an original research article published in the *Journal of Theoretical Biology*, Volume 65, Issue 1, March 7, 1977, pp. 29–77, © Elsevier.

5) "An Application of the Contact Pressure Model of Phyllotaxis to the Close Packing of Spheres around a Cylinder in Biological Fine Structure" is an original research article published in the *Journal of Theoretical Biology*, Volume 67, Issue 3, August 7, 1977, pp. 447–458, © Elsevier,

6) "The Role of Continued Fractions in Phyllotaxis" was originally published in the *Journal of Algebra*, Volume 205, Issue 1, July 1, 1998, pp. 227–243, © Elsevier

7) "The Role of Mathematics in Phyllotaxis" is the Prologue for the book *Symmetry in Plants*, © World Scientific and published in 1998.

8) "Generating Phyllotaxis Patterns on a Cylindrical Point Lattice" is Chapter 11 of the book *Symmetry in Plants*, © World Scientific and published in 1998.

J. theor. Biol. (1974) **45,** 1–79

A Model of Contact Pressure in Phyllotaxis

IRVING ADLER

North Bennington, Vermont 05257, U.S.A.

(*Received* 24 *May* 1972, *and in revised form* 7 *November* 1973)

> "*Wie es nun eine Tendenz gibt, vermöge welcher die Blätter sich am unabhängigsten darzu bilden, sich möglichst von einander entfernen und isolieren, so liegt auch noch eine andere Nothwendigkeit in der Erzeugung der meisten, vermöge welcher sie einander mehr genähert sind.*"
> Karl Friedrich Schimper, 1830

Using a cylindrical representation of a leaf distribution with equal internode distances and equal successive divergences, the precise connection between divergence and parastichy numbers is determined. Then, using the assumption that contact pressure maximizes the minimum distance between leaf primordia, a testable model of contact pressure is constructed. A particular condition on the ratio of the internode distance to the girth of the stem is identified that suffices for contact pressure to restrict the divergence to the values that produce normal Fibonacci phyllotaxis with parastichy numbers that are consecutive terms of the sequence 1, 2, 3, 5, ... It is argued that a contact pressure model is relevant because a Richards-type inhibitor field theory alone does not suffice to explain this restriction.

Contents

1. Some Parts of the Problem Are Mathematical

The problem of phyllotaxis is partly biological and partly mathematical. Determining by observation what geometric configuration best represents the arrangement of leaves around a stem is a biological problem. Exploring the geometry of the configuration is a mathematical problem. Selecting appropriate hypotheses to portray biological processes that might explain the genesis of observed configurations is a biological problem. Determining whether certain hypotheses do or do not imply particular conclusions is a mathematical problem. This paper deals with some parts of the problem of phyllotaxis that are purely mathematical. They will be stated explicitly in section 5, after facts, concepts and terminology that are needed to express them are introduced.

2. Preliminary Definitions and Facts about Fibonacci Numbers

A sequence (a_n), namely, $a_1, a_2, \ldots, a_n, \ldots$, is called a Fibonacci sequence if it satisfies the recurrence relation

$$a_{n+1} = a_n + a_{n-1}, \qquad n > 1. \tag{1}$$

Each Fibonacci sequence is completely determined by its first two terms. Two types of Fibonacci sequence play a significant role in phyllotaxis. In one type, which plays a major role, the first two terms of the sequence are 1 and t respectively, where t is a positive integer. In the second type, which plays a minor role, the first two terms of the sequence are t and $2t+1$ respectively. In the special case $t = 1$ of the first type of sequence, it is customary to use the notation F_n for the nth term of the sequence. Thus,

$$(F_n) = 1, 1, 2, 3, 5, 8, 13, 21, 34, \ldots \tag{2}$$

This sequence, since it was the first Fibonacci sequence to receive attention in mathematical literature, is often referred to as *the* Fibonacci sequence. For $t > 1$, the notation $S_{t,n}$ is used for the nth term of the sequence of the first type for given t. However, in order to simplify some formulas that occur later, the terms of $(S_{t,n})$ are indexed starting with 0, i.e.,

$$S_{t,0} = 1, \qquad S_{t,1} = t,$$

and

$$(S_{t,n}) = 1, t, t+1, 2t+1, 3t+2, \dots \tag{3}$$

Sequence (3) is related to sequence (2) by

$$S_{2,n} = F_{n+2}. \tag{4}$$

$$S_{t,n} = F_n t + F_{n-1}. \tag{5}$$

Sequence (F_n) is related to the golden section τ, which is the positive root of the equation $x^2 - x - 1 = 0$:

$$\lim_{n \to \infty} \frac{F_{n+1}}{F_n} = \tau = \frac{1+\sqrt{5}}{2} \doteq 1 \cdot 62. \tag{6}$$

$$\lim_{n \to \infty} \frac{F_{n-1}}{F_{n+1}} = 1 - \tau^{-1} = \tau^{-2} = \frac{3-\sqrt{5}}{2} \doteq 0 \cdot 38. \tag{7}$$

Sequences (F_n) and $(S_{t,n})$ and the golden section τ are all related to the non-terminating periodic continued fraction in which the first term under the fraction line is t and all others are 1:

$$\cfrac{1}{t + \cfrac{1}{1 + \cfrac{1}{1 + \dots}}} = \frac{1}{t + \tau^{-1}}. \tag{8}$$

The convergents of the continued fraction (8) are

$$\frac{F_n}{S_{t,n}}. \tag{9}$$

In the special cases $t = 2, 3$ and 4, which are particularly important in phyllotaxis, the continued fraction (8) has these values:

$$\frac{1}{2+\tau^{-1}} = \tau^{-2} \doteq 0 \cdot 38,$$

$$\frac{1}{3+\tau^{-1}} \doteq 0 \cdot 28, \tag{10}$$

$$\frac{1}{4+\tau^{-1}} \doteq 0 \cdot 22.$$

Sequence 1, 3, 4, 7, ... is sometimes referred to in mathematical literature as the Lucas sequence, and its nth term is designated as L_n. In our notation, $L_n = S_{3,n-1}$.

The terms F_n of sequence (2) and the terms $S_{t,n}$ of sequence (3) satisfy many identities that can be derived from the recurrence relation (1) and the values of the first two terms of each sequence (Hoggatt, 1969). Some identities that will be used in this paper are listed in Appendix A.

3. Preliminary Definitions and Facts about Phyllotaxis

Phyllotaxis is the study of the arrangement of leaves on a stem, scales on a pine cone, florets on a pineapple, florets on a sunflower, etc. These arrangements are found on many types of surface. For example, leaves on a mature stem and florets on a pineapple are distributed on a surface that is approximately cylindrical. Leaf primordia on the growing tip of a stem are on a surface that is approximately conical. The florets of a sunflower are on a surface that is almost a disc (the interior of a circle). The facts of phyllotaxis can be described most simply if the distribution is represented as being on a cylindrical surface. Moreover, a representation of the distribution on another kind of surface can be converted into a cylindrical representation by an appropriate geometric transformation. So in what follows we represent the distribution as being on a cylindrical surface with vertical axis.

We shall be concerned only with the regions on the surface where the leaves, scales, florets, etc., are attached. Any such region will be referred to as a "leaf". It is important to keep this convention in mind, so that it will be understood that statements about leaves are statements about regions on a surface, and not about appendages attached at these regions. For some purposes it will suffice to picture a leaf as a point, for others as a circular region.

There may be one or more leaves at a given level (node) on the surface. We are concerned in this paper only with the most common case of one leaf per node. However, the results of this paper are easily generalized to include the case of more than one leaf node. How this can be done will be indicated in section 14.

If each leaf is joined to the next higher one by the shortest path on the surface, a curve is obtained that is called the fundamental spiral. We assume in most of what follows that the fundamental spiral is a helix, and that the leaves are at equal intervals on it, so that the centers of the leaves form a cylindrical lattice. The leaves are numbered in order on the fundamental spiral and occasionally any arbitrary leaf is used as leaf 0. At other times we shall denote by 0 the first leaf to emerge on the stem.

The following notations are used for the indicated variables that are needed to describe a leaf distribution:

c = girth of the cylinder;

h_n = height of leaf n above the level of leaf 0;

l = internode distance = $h_n - h_{n-1}$ for all n;

Δ = leaf diameter;

d = divergence of each leaf n from leaf $n-1$

= the fraction of a turn between them around the axis of the cylinder, assumed to be the same for all n; $(d \leqslant 1/2)$;

plastochrone

= the interval of time between the emergence of two successive leaf primordia on the growing tip of a stem.

t_n = the time elapsed between the emergence of leaf 0 and the emergence of leaf n.

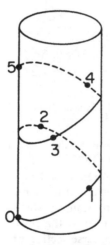

FIG. 1. Leaves on one fundamental spiral with $d = 2/5$.

Most of the time we shall be using a normalized representation of a leaf distribution in which the girth of the cylinder is taken as the unit of length, and the plastochrone is taken as the unit of time. In the normalized representation elapsed time is denoted by T, and the time of emergence of leaf n by T_n, with $T_0 = 0$. Note that $T_n = n$. The internode distance will be called the *rise*, r, and the leaf diameter the *relative leaf diameter*, δ. These quantities are related to the corresponding quantities for a non-normalized picture by the following equations:

$$r = l/c, \qquad \delta = \Delta/c, \qquad T_n = t_n/\text{plastochrone}. \qquad (11)$$

The measure δ is related to the "bulk ratio" of Church by the relation: bulk ratio $= \sin \pi\delta$ (Richards, 1948, p. 226).

In any leaf distribution, there are secondary spirals (parastichies) determined by joining a leaf to any other leaf. These may spiral up to the left or to the right. Parastichies occur in parallel sets that effect a partition of the entire leaf distribution, so that every leaf lies on one and only one member of the set. A complete set of left parastichies (going up to the left) with m members and a complete set of right parastichies (going up to the right) with n members is called the *opposed parastichy pair* (m, n).

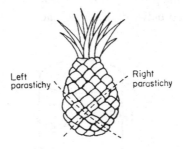

Left parastichy

Right parastichy

FIG. 2. Parastichies in a pineapple.

The investigators who initiated the systematic study of phyllotaxis, viz. Schimper (1830) and Braun (1831, 1835) and the Bravais brothers (1837), used a cylindrical representation of a leaf distribution because it pictured approximately what is seen on a mature stem. Church (1904) introduced what he called the centric representation, in which the leaves are distributed on a disc (with the oldest leaves nearest the rim), because this pictures approximately what is seen in a transverse section of the growing tip of a stem. The significant variables in a centric representation are:

d = divergence of each leaf from its predecessor
= the fraction of a turn between them around the center of the disc;
ρ_n = the distance of leaf n from the center of the disc;
R = the plastochrone ratio = ρ_{n-1}/ρ_n for all n.

If (x, y) are rectangular co-ordinates for a point in the plane development of a normalized cylindrical representation, and (ρ, θ) are polar co-ordinates for the corresponding point in a centric representation, the transformation defined by

$$\rho = (2\pi)^{-1} e^{-2\pi y}, \qquad \theta = 2\pi x \qquad (12)$$

maps one into the other so that heights y that are in arithmetic progression correspond to radii ρ that are in geometric progression, and orthogonality of opposed parastichy pairs is preserved. Under this transformation, the plasto-chrone ratio R and the rise r are related by the formula

$$R = e^{2\pi r}. \qquad (12')$$

By means of this formula, predictions made by the mathematical model developed here can be converted into their centric equivalents so that they may be checked against observational data. (Tests of the model are discussed in section 13.)

In an arbitrary opposed parastichy pair there need not be a leaf at every point of intersection of a left parastichy and a right parastichy. In the special case where there is a leaf at every intersection, the opposed parastichy pair is called *visible*.

If (m, n) is a visible opposed parastichy pair, there is exactly one funda-mental spiral if and only if m and n are relatively prime (Bravais, 1837, p. 54).

In Appendix B it can be seen that in a leaf distribution with infinitely many leaves there are infinitely many visible opposed parastichy pairs. However, usually only one and occasionally two of them are conspicuous. A *conspicuous* visible opposed parastichy pair is determined by a leaf and its nearest neighbors to the right and to the left. The contact parastichies that are discussed so frequently in the contemporary literature on phyllotaxis are conspicuous opposed parastichy pairs. If the conspicuous opposed parastichy pair on a plant is (m, n) it is customary to say that the plant has (m, n) phyllotaxis.

4. The Principal Facts of Phyllotaxis (with One Fundamental Spiral)

The principal observed facts of phyllotaxis may be summed up as follows:

(1) If leaves are numbered in the order of their emergence, and if d_n is the divergence of leaf n from leaf $n-1$, then the initial values of the first terms of the sequence (d_n) differ from each other, but the sequence converges rapidly to a limit d as n increases; meanwhile, the first terms of the sequence change from their initial values with the passage of time and also approach d as a limit (Richards, 1948, p. 218). Consequently leaves are soon arranged at approximately equal intervals on a helix.

(2) In nearly all cases, measured values of d (rational approximations) have the form

$$d = \frac{F_h}{S_{t,h}} \qquad (13)$$

which are approximations to $(t+\tau^{-1})^{-1}$ (Bravais, 1837), and if (m, n) is a conspicuous opposed parastichy pair, m and n are consecutive terms of $S_{t,h}$ for some value of t. This most common type of phyllotaxis will be called *normal* phyllotaxis. In about 95% of the cases of normal phyllotaxis $t = 2$ (Bravais, 1837, p. 74), and then

$$d = \frac{F_h}{F_{h+2}},\tag{14}$$

which are approximations to τ^{-2}, and m and n are consecutive terms of F_h. During the period in which the characteristic phyllotaxis of a plant is formed, the index h increases from 1 to a value that is characteristic of the species.

(3) In nearly all cases of normal phyllotaxis, values of t are restricted to 2, 3 and 4. That is,

$$d = (2+\tau^{-1})^{-1}, \text{ or } (3+\tau^{-1})^{-1}, \text{ or } (4+\tau^{-1})^{-1};\tag{15}$$

or

$$d^{-1}-\tau^{-1} \doteq 2, 3 \text{ or } 4 \text{ (Wright, 1873, p. 385).}\tag{16}$$

However, cases of normal phyllotaxis corresponding to other values of t have been observed (Fujita, 1937, p. 488).

(4) Among the rare cases of phyllotaxis that are not normal, the most frequent type (m, n) is one in which m and n are consecutive terms of a Fibonacci sequence whose first two terms have the form $t, 2t+1$ with $t \geqslant 2$. We shall call this type of phyllotaxis *anomalous*. In the case $t = 2$ of anomalous phyllotaxis, m and n are consecutive terms of the sequence

$$2, 5, 7, 12, 19, \ldots.\tag{17}$$

5. Purpose

In this paper the following questions are dealt with:

(1) What is the precise connection between parastichy numbers and divergence in a cylindrical lattice?

(2) Assuming that there is a tendency (whose cause is not specified) for the leaf distribution of a plant to be approximately a regular cylindrical lattice (in the cylindrical representation), how is this tendency modified when contact pressure is present? In the discussion of this question we shall take up the following sub-questions: (a) What mathematical assumption about a cylindrical lattice corresponds to the biological assumption of the presence of contact pressure? (b) What are the general consequences of this assumption? (c) What general method may be used to determine whether, for a given species of plant, contact pressure suffices to explain the observed changes in phyllotaxis? (d) Under the simplifying assumption that a leaf distribution is a regular cylindrical lattice, what are conditions under which contact pressure

would produce normal Fibonacci phyllotaxis? What are conditions under which contact pressure would produce anomalous phyllotaxis?

(3) What role can a field theory of the type described by Richards (1948) play in the theory of phyllotaxis? In particular, can such a field theory *alone* explain the genesis of normal phyllotaxis? Can an appropriate field theory, *combined with a contact pressure theory*, as originally proposed by Schoute (1913), explain the principal facts of phyllotaxis?

6. How the Divergence Is Related to the Visible Opposed Parastichy Pairs

The statement of fact (2) in section 4 indicates that there is an observed linkage between the divergence of the leaf distribution on a plant and the parastichy numbers that are observed on it. That there is such a linkage was known from the very beginning of the systematic study of phyllotaxis. Braun (1835) referred to parastichies as "diagnostic rows" from whose parastichy numbers divergences assumed to be rational could be calculated when they were not directly measurable. The Bravais brothers (1837, p. 53) calculated an upper and lower bound for the divergence (rational or irrational) from the parastichy numbers m and n of an observed opposed parastichy pair (m, n). Richards (1951, p. 522) pointed out that to each particular phyllotaxis (conspicuous opposed parastichy pair in our terminology) there corresponds a range of possible divergences, and the higher the parastichy numbers, the smaller the range. However, underlying the connection between divergence and a conspicuous opposed parastichy pair is a connection between divergence and visible opposed parastichy pairs. We pause now to examine this connection, which will supply the answer to question (1) of section 5.

The connection between the divergence of a leaf distribution and its visible opposed parastichy pairs depends on the position of the leaves but not on the size of the leaves. The problem of finding this connection is therefore a purely geometric problem in the study of a cylindrical lattice. Since no complete and rigorous solution of the problem has been published before, the sequence of propositions by which it is solved is outlined in Appendix B. The proofs of the propositions are all elementary and are omitted. In what follows in this section we limit ourselves to defining the concepts on which the solution of the problem depends, and stating as simply as possible the results that will be used.

The principal concepts and theorems that are involved in the solution of the problem may be summarized as follows:

(1) The concepts of *visible* opposed parastichy pair and *conspicuous* opposed parastichy pair are defined and distinguished from each other as in section 3.

(2) Using any arbitrary leaf as leaf 0, the leaves are numbered according to the order of their positions on the fundamental spiral.

(3) The cylindrical surface is slit along the element of the cylinder that passes through point 0, and is then unrolled onto a plane, with the point 0 on the left, and its duplicate, designated as 0_1, on the right, so that the entire cylindrical surface is mapped into the strip of the plane between parallel lines passing through 0 and 0_1. Since we are using the normalized representation on a cylinder, the length of line segment 00_1 is 1.

(4) If (m, n) is a visible opposed parastichy pair, one of the m left parastichies goes up to the left from 0_1, and one of the n right parastichies goes up to the right from 0. These two lines and the line 00_1 determine a triangle with the following properties: The base of the triangle has length 1. The first leaf above 0 on the left leg is leaf n. The first leaf above 0_1 on the right leg is leaf m. If the line segment between two consecutive leaves on each leg of the triangle is called a *step*, there are m steps on the left leg, and n steps on the right leg. There is a leaf at the apex of the triangle, and in fact it is leaf number mn (see Fig. 3). This triangle is called the visible opposed parastichy triangle belonging to the visible opposed parastichy pair (m, n). There is a one-to-one correspondence between visible opposed parastichy pairs and visible opposed parastichy triangles. This makes it possible to show that an opposed parastichy pair is visible by showing that it has a visible opposed parastichy triangle.

(5) If (m, n) is a visible opposed parastichy pair, with $m > n$, there is a visible opposed parastichy pair $(m-n, n)$. If $n > m$, there is a visible opposed parastichy pair $(m, n-m)$. $(m-n, n)$ or $(m, n-m)$, whichever the case may be, is called the *contraction* of (m, n).

(6) If (m, n) is a visible opposed parastichy pair, the parastichy pair $(m+n, n)$ is called its *left extension*, and the parastichy pair $(m, m+n)$ is called its *right extension*. An extension of a visible opposed parastichy pair need not be a visible opposed parastichy pair. The central problem of Appendix B is to determine the conditions under which a left or right extension of a visible opposed parastichy pair is a visible opposed parastichy pair.

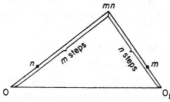

FIG. 3. Visible opposed parastichy triangle for the visible opposed parastichy pair (m, n).

(7) To solve this problem, it is shown first that, if the fundamental spiral is a right spiral, every visible opposed parastichy pair (m, n) with $m, n > 1$ can be obtained as the end product of a sequence of extensions starting with a visible opposed parastichy pair of the form $(t, t+1)$, where t is a uniquely determined integer greater than 1. It is also shown that $(t, t+1)$ is a visible opposed parastichy pair if and only if $1/(t+1) \leqslant d \leqslant 1/t$, that is, if and only if the number d is in the closed segment $[1/(t+1), 1/t]$ on the real number line.

(8) The next step in the argument requires the concept of the mediant between two fractions a/b and c/d that are in lowest terms. Their mediant is the fraction $(a+c)/(b+d)$. The mediant between $1/(t+1)$ and $1/t$ is the fraction $2/(2t+1)$. It divides the segment $[1/(t+1), 1/t]$ into two segments, a left-hand segment lying to the left of the mediant, and a right-hand segment lying to the right of the mediant. It is proved that the left extension $(2t+1, t+1)$ of $(t, t+1)$ is visible if and only if d lies in the left-hand segment $[1/(t+1), 2/(2t+1)]$, and the right extension $(t, 2t+1)$ of $(t, t+1)$ is visible if and only if d lies in the right-hand segment $[2/(2t+1), 1/t]$. Moreover, this procedure can be repeated with each of these extensions. For example, start now with the fact that $(2t+1, t+1)$ is a visible opposed parastichy pair if and only if d is in the closed segment $[1/(t+1), 2/(2t+1)]$. The mediant between the endpoints of this segment is $3/(3t+2)$, and it divides the segment into two segments. The left extension $(3t+2, t+1)$ of $(2t+1, t+1)$ is a visible opposed parastichy pair if and only if d lies in the left-hand segment $[1/(t+1), 3/(3t+2)]$, and the right extension $(2t+1, 3t+2)$ is a visible opposed parastichy pair if and only if d lies in the right-hand segment $[3/(3t+2), 2/(2t+1)]$, and so on. This is the meaning of Theorem 1 in Appendix B.

(9) Theorem 2 sums up the result of using the procedure described above k times in succession. The end product of k successive extensions starting with $(t, t+1)$ is called an extension of $(t, t+1)$ of order k. The left extension and the right extension of $(t, t+1)$ are each of order 1. Let us designate them as L and R respectively. The left extension L has both a left and right extension which are designated LL and LR respectively. The right extension R has both a left and right extension labeled RL and RR, respectively. Thus there are four extensions of $(t, t+1)$ of order 2, and their designations LL, LR, RL and RR, when read from left to right tell us how they are obtained from $(t, t+1)$. For example, LR is obtained by taking first a left extension of $(t, t+1)$, namely $(2t+1, t+1)$, and then taking a right extension of that to obtain $(2t+1, 3t+2)$. Similarly, there are eight extensions of order 3, designated as LLL, LLR, etc.

We introduce an analogous notation for successive subdivisions of the segment $[1/(t+1), 1/t]$ obtained by repeatedly inserting mediants. Inserting $2/(2t+1)$, the mediant between the endpoints of the segment, gives two segments designated as L and R: L for the left-hand segment and R for the right-hand segment. This gives the median subdivision of $[1/(t+1), 1/t]$ of order 1. Now by inserting in each of these segments the median between its endpoints, the median subdivision of $[1/(t+1), 1/t]$ of order 2 is obtained. It consists of four segments designated as LL, LR, RL and RR respectively. The designation, when read from left to right, indicates how the segment is obtained from $[1/(t+1), 1/t]$. The segment LR is obtained by taking the left segment of $[1/(t+1), 1/t]$, namely $[1/(t+1), 2/(2t+1)]$ and then taking the right segment of that, namely $[3/(3t+2), 2/(2t+1)]$. Theorem 2 asserts that extension LR of $(t, t+1)$ is a visible opposed parastichy pair if and only if the divergence d lies in segment LR of $[1/(t+1), 1/t]$; extension LL is visible if and only if d lies in segment LL; extension RL is visible if and only if d lies in segment RL; and extension RR is visible if and only if d lies in segment RR. Similarly, with extensions of order 3 and the median subdivision of order 3, extension LRL is visible if and only if d lies in segment LRL, etc.

(10) The special case of Theorem 2 that is of particular interest in connection with the problem of phyllotaxis is the case in which the successive extensions of $(t, t+1)$ are alternately left and right, starting with a left extension of $(t, t+1)$. This special case is covered by Theorem 4. Expressed in terms of $h = k+1$, Theorem 4 indicates that if (m, n) is a visible opposed parastichy pair, then m and n are consecutive terms $S_{t,h}$ and $S_{t,h+1}$ of $(S_{t,h})$ if and only if d is either $F_h/S_{t,h}$ or $F_{h+1}/S_{t,h+1}$ or lies between them.

In the important case $t = 2$, Theorem 4 gives the following sequence of statements concerning successive extensions of $(2, 3)$ that are alternately left and right. In these statements the values of $F_h/S_{2,h} = F_h/F_{h+2}$ are given explicitly:

(2, 3) is visible if and only if $1/3 \leqslant d \leqslant 1/2$;
(5, 3) is visible if and only if $1/3 \leqslant d \leqslant 2/5$;
(5, 8) is visible if and only if $3/8 \leqslant d \leqslant 2/5$;
(13, 8) is visible if and only if $3/8 \leqslant d \leqslant 5/13$;

etc.

These statements tell us just how the visible opposed parastichy pairs are related to the divergence in normal phyllotaxis with $t = 2$.

The theorem also gives the following statements concerning successive extensions of $(2, 3)$ that are alternately right and left:

(2, 3) is visible if and only if $1/3 \leqslant d \leqslant 1/2$;
(2, 5) is visible if and only if $2/5 \leqslant d \leqslant 1/2$;
(7, 5) is visible if and only if $2/5 \leqslant d \leqslant 3/7$;
(7, 12) is visible if and only if $5/12 \leqslant d \leqslant 3/7$;

etc.

These statements tell us how visible opposed parastichy pairs are related to the divergence in anomalous phyllotaxis with $t = 2$.

(11) If all the (infinitely many) extensions of $(t, t+1)$ obtained by taking alternately left and right extensions are visible, then $d = 1/(t+\tau^{-1})$, and conversely.

(12) For any given leaf distribution, there exists either exactly one or exactly two infinite sequences of visible opposed parastichy pairs, each an extension of the one that precedes it, according to whether d is irrational or rational. (For any given divergence d, the infinite sequence or sequences of visible opposed parastichy pairs that it determines can be identified from the representation of d as a simple continued fraction. This will be the subject of a separate paper.) By (11) above, such an infinite sequence of successive extensions will be alternately left and right extensions of $(t, t+1)$ only if $d = 1/(t+\tau^{-1})$.

(13) In a normalized representation of a leaf distribution with rise r, let l_0 be the element of the cylinder through leaf 0, and write dist (l_0, a) for the distance between l_0 and a. Then dist (l_0, a) is the horizontal component of the distance from leaf 0 to leaf a, and ra is the vertical component. If we write dist $(0, a)$ for the distance from 0 to a, we have

$$\text{dist } (0, a) = [\text{dist}^2 (l_0, a) + (ar)^2]^{\frac{1}{2}}. \tag{18}$$

Suppose that for a given value of r leaf numbers a and b are such that leaf a is nearer to leaf 0 than leaf b is, $b > a$, but dist $(l_0, b) <$ dist (l_0, a). Then, if r is allowed to decrease sufficiently, leaf b will ultimately be nearer to leaf 0 than leaf a. For this reason, as r is decreased, a conspicuous opposed parastichy pair ceases to be conspicuous and is displaced in its role of conspicuous opposed parastichy pair by one with higher parastichy numbers. Only a visible opposed parastichy pair can become a conspicuous opposed parastichy pair. To assume that with falling r the succession of conspicuous opposed parastichy pairs is precisely the sequence of successive extensions of $(t, t+1)$ obtained by taking alternately left and right extensions is tantamount to assuming that the divergence is $1/(t+\tau^{-1})$. This is important to know in order to avoid the trap of circular reasoning in any attempt to explain the occurrence of this divergence. For example, if $(2, 3)$ is conspicuous, $1/3 \leqslant d \leqslant 1/2$. We may not assume that as r falls $(2, 3)$ will be succeeded in the role of conspicuous pair by $(5, 3)$. This will happen only if $1/3 \leqslant d \leqslant 2/5$. On the other hand, if $2/5 \leqslant d \leqslant 1/2$, $(2, 3)$ will be succeeded by $(2, 5)$, and the phyllotaxis would be revealed as anomalous rather than normal.

7. Simplifying Assumptions and Their Purposes

(a) In observed leaf distributions, successive divergences d_n are not all equal, but tend toward approximate equality with increasing n. We shall make the simplifying assumption that they are all equal, with some value d. (Without loss of generality, it is also assumed that the fundamental spiral is a right spiral.) This will make it possible to determine the underlying principle of the influence of contact pressure on divergence. Once the principle is known, it will help illuminate the problem of calculating the effect of contact pressure when successive d_n are not precisely equal. (The latter problem is discussed in section 13.)

(b) It is assumed that the leaves in the cylindrical representation of a leaf distribution are congruent regions. In order to have a simple and convenient measure for the size of a leaf, it is assumed that these regions are circular. This is an inessential assumption entailing no loss in generality. (This matter is discussed further at the end of section 11.)

(c) It is assumed that during the initial period of growth of the stem apex and emergence of new leaf primordia the girth c of the stem grows faster than the internode distance l, so that $r = l/c$ decreases. (The reason for this assumption is explained at the end of section 8.)

(d) Contact pressure arises when each leaf grows in diameter until it touches its nearest neighbor and then tends to grow larger after that. We shall make the simplifying assumption that when there is contact pressure the relative leaf diameter $\delta = \Delta/c$ is maximized. (See sections 8 and 9 for the largest possible value of δ, and its determination.)

The analysis of the consequences of contact pressure is developed in sections 8, 9, 10, 11 and 12, and is divided as follows. In section 8 the consequences of maximation of δ is discussed when d is fixed. The need for, and definition of, the concept "points of close return" is established. Some theorems about these points in the case of normal phyllotaxis are proven, and these theorems used to derive a formula that relates the phyllotaxis of a plant to its rise r. In section 9 the consequences of the maximization of δ if d is allowed to vary are examined. In section 10 special conditions are identified under which the maximization of δ produces normal phyllotaxis and in section 11 special conditions are identified under which the maximization of δ produces anomalous phyllotaxis.

8. Maximum Relative Leaf Diameter for Fixed Divergence

The leaf centers form a cylindrical lattice. It is obvious that for an arbitrary lattice with fixed d and r, there is a maximum possible relative leaf diameter δ equal to the *minimum distance between leaf centers*. If x and y are any two leaf numbers, with $y > x$, dist $(x, y) = $ dist $(0, y-x)$, because of the symmetry of a cylindrical lattice. It follows then that

$$\max \delta = \min \text{dist}(0, x), \quad x \geqslant 1, (d \text{ and } r \text{ fixed}). \tag{19}$$

Let a be the value of x for which dist $(0, x)$ is a minimum. Now let us consider what happens as r decreases. We saw in part (13) of section 6 that, if r is small enough, leaf a may be displaced in its role of nearest neighbor to leaf 0 by some leaf b with $b > a$. Exploring this possibility more closely, the horizontal components of dist $(0, a)$ and dist $(0, b)$ are dist (l_0, a) and dist (l_0, b) respectively. The vertical components are ar and br respectively. If $b > a$, $br > ar$. If dist $(l_0, b) \geqslant$ dis (l_0, a), then it will follow that dist $(0, b) >$ dist $(0, a)$. Consequently leaf b can be closer to leaf 0 than leaf a (if r is sufficiently small) only if dist $(l_0, b) <$ dist (l_0, a). The first leaf after leaf a that has this property will be the first leaf capable of displacing a as nearest neighbor to leaf 0, as r decreases. This is the motivation for introducing the concept of *points of close return*, defined as follows:

Definition

Let l_0 be the element of the cylindrical surface that passes through leaf 0. Let $n_1 = 1$. Let n_2 be the first leaf after leaf 1 that is closer to l_0 than n_1 is, but is not on it. Let n_3 be the first leaf after n_2 that is closer to l_0 than n_2 is, but is not on it; and so on. The leaves $n_1, n_2, n_3, \ldots$ are called the *points of close return*.

If a leaf n is not on l_0, the distance of leaf n from l_0 measured to the right is the fractional part of nd. Let us denote this measure by $d(n)$. Let $c(n) = 1 - d(n) =$ the distance of leaf n from l_0 measured to the left. For $i > 1$, if n_i exists, dist (l_0, n_i) is equal to either $d(n_i)$ or $c(n_i)$, whichever is smaller. In the first case we say n_i is to the right of l_0. In the second case we say n_i is to the left of l_0. It is not difficult to show that any two consecutive points of close return are on opposite sides of l_0. When the fundamental spiral goes up to the right, n_k is to the right of l_0 if k is odd, and to the left if k is even. For brevity the notation $D_i = $ dist (l_0, n_i) is used.

For any fixed d, all the points of close return are well-defined. There are finitely many if d is rational, and infinitely many if d is irrational. In what follows, however, we shall be concerned not with a particular value of d but with all the values of d in a particular interval. For such a range of values of d at least some of the points of close return may be well-defined. To illustrate this fact, three specific ranges of values of d are considered that will be of interest to us later. Considering first the interval $1/3 < d < 1/2$:

$$(1) \qquad \begin{array}{c|c|c} 1/3 < d < 1/2 & 1/3 < d(1) < 1/2 & 1/2 < c(1) < 2/3 \\ 2/3 < 2d < 1 & 2/3 < d(2) < 1 & 0 < c(2) < 1/3 \\ 1 < 3d < 3/2 & 0 < d(3) < 1/2 & 1/2 < c(3) < 1. \end{array}$$

Since $c(2) < d(2)$, dist $(l_0, 2) = c(2)$. Since $c(2) < d(1)$, $n_2 = 2$, and $D_2 = c(2) = 1 - 2d$. Since $d(3) < c(3)$, dist $(l_0, 3) = d(3)$. $n_3 = 3$ if and only if $d(3) < c(2)$. Since $d(3) = 3d - 1$, and $c(2) = 1 - 2d$, the condition for n_3 to be equal to 3 is $3d-1 < 1-2d$, or $d < 2/5$. That is, $n_3 = 3$ only in part of the interval $1/3 < d < 1/2$, namely, the interval $1/3 < d < 2/5$. Let us consider this smaller interval next:

(2)

$1/3 < d < 2/5$	$1/3 < d(1) < 2/5$	$3/5 < c(1) < 2/3$
$2/3 < 2d < 4/5$	$2/3 < d(2) < 4/5$	$1/5 < c(2) < 1/3$
$1 < 3d < 6/5$	$0 < d(3) < 1/5$	$4/5 < c(3) < 1$
$4/3 < 4d < 8/5$	$1/3 < d(4) < 3/5$	$2/5 < c(4) < 2/3$
$5/3 < 5d < 2$	$2/3 < d(5) < 1$	$0 < c(5) < 1/3$

As before, $n_2 = 2$, and as anticipated above, $n_3 = 3$. Then $D_3 = d(3) = 3d-1$. Notice that leaf 4 cannot possibly be n_4, because both $d(4)$ and $c(4)$ are greater than $d(3)$. However, leaf 5 is not excluded altogether from being n_4, because $c(5) < d(3)$ for an appropriate subinterval of the interval under consideration. In fact, since $c(5) = 2-5d$, and $d(3) = 3d-1$, $n_4 = 5$ if and only if $2-5d < 3d-1$, or $d > 3/8$.

The intervals $1/3 < d < 1/2$, $1/3 < d < 2/5$, and $3/8 < d < 2/5$ are ranges of values of d associated with $(2, 3)$, $(5, 3)$ and $(5, 8)$ phyllotaxis respectively, all examples of normal phyllotaxis. These examples illustrate that the smaller the range of values of d, the more points of close return can be identified for the whole range. In fact we shall derive a general rule which specifies which are the identifiable points of close return for all the intervals of values of d cited in Theorem 4 as being associated with normal phyllotaxis. However, before doing so, one more specific interval of a different type is considered, namely $2/5 < d < 1/2$, associated with anomalous phyllotaxis.

(3)

$2/5 < d < 1/2$	$2/5 < d(1) < 1/2$	$1/2 < c(1) < 3/5$
$4/5 < 2d < 1$	$4/5 < d(2) < 1$	$0 < c(2) < 1/5$
$6/5 < 3d < 3/2$	$1/5 < d(3) < 1/2$	$1/2 < c(3) < 4/5$
$8/5 < 4d < 2$	$3/5 < d(4) < 1$	$0 < c(4) < 2/5$
$2 < 5d < 5/2$	$0 < d(5) < 1/2$	$1/2 < c(5) < 1$

As before, $n_2 = 2$. But now n_3 is not 3, because both $d(3)$ and $c(3)$ are greater than $c(2)$. Because the range of possible values of $c(4)$ is between 0 and $2/5$, it looks as though n_3 might be 4 in an appropriate subinterval. But this appearance is deceptive for the following reason. It can be seen from the table above that $c(4) = 2-4d$, and $c(2) = 1-2d$. Therefore $c(4) = 2[c(2)] > c(2)$, and n_3 cannot possibly be 4, although, it may be 5. In fact, $n_3 = 5$ if and only if $d(5) < c(2)$. But $d(5) = 5d-2$, and $c(2) = 1-2d$. Therefore, the condition for n_3 to be 5 is $5d-2 < 1-2d$, or $d < 3/7$.

Examples (2) and (3) introduce another concept that will be useful in what follows. In both cases, although it was not possible to identify a leaf after n_2 that could be n_3 for the whole interval, it was possible to identify which leaf after n_2 could be n_3 for a smaller interval within it. This observation leads to the following definition:

If n_i is a point of close return for some interval I of values of d, and $n > n_i$ has the property that (a) $n = n_{i+1}$ for a subinterval of I, and (b) for $n_i < b < n$, dist $(l_0, b) \geqslant$ dist (l_0, n_i) throughout I, then n is called the *next close return candidate after* n_i. The notation n'_{i+1} shall be used for the next close return candidate after n_i, and D_{i+1} for dist (l_0, n'_{i+1}).

We are now ready to state a general theorem about points of close return for the intervals of values of d associated with normal phyllotaxis.

THEOREM 7

(A) *If*

$$F_{2k}/S_{t, 2k} < d < F_{2k-1}/S_{t, 2k-1} \text{ for } k \geqslant 1,$$

then

$$n_i = S_{t, i-1} \text{ for } i = 1, \ldots, 2k,$$

and

$$n'_{2k+1} = S_{t, 2k}.$$

For

$$i = 1, \ldots, 2k+1, D_i = (-1)^i(F_{i-1} - S_{t, i-1}d).$$

(B) *If*

$$F_{2k}/S_{t, 2k} < d < F_{2k+1}/S_{t, 2k+1} \text{ for } k \geqslant 1,$$

then

$$n_i = S_{t, i-1} \text{ for } i = 1, \ldots, 2k+1,$$

and

$$n'_{2k+2} = S_{t, 2k+1}.$$

For

$$i = 1, \ldots, 2k+2, D_i = (-1)^i(F_{i-1} - S_{t, i-1}d).$$

The proof is by induction, and is based on repeated use of the following: (1) $S_{t,h}$ and $S_{t,h+1}$ are relatively prime; (2) if $d(a) + d(b) < 1$, then $d(a+b) = d(a) + d(b)$; (3) identities A1, A3, A4 and A5 of Appendix A; (4) the well-known theorem that if r and s are relatively prime integers, and x is any integer, the least residues modulo s of x, $x+r$, $x+2r$, $\ldots$, $x+(s-1)r$ are 0, 1, $\ldots$, $s-1$ rearranged (Le Veque, 1956, p. 27).

The two parts of the theorem can be combined into a single statement as follows: If d is between $F_h/S_{t,h}$ and $F_{h-1}/S_{t,h-1}$ with $h \geqslant 1$, then

$$n_i = S_{t, i-1} \text{ for } i = 1, \ldots, h;$$

$$n'_{h+1} = S_{t, h};$$

and

$$D_i = (-1)^i(F_{i-1} - S_{t, i-1}d) \text{ for } i = 1, \ldots, h+1.$$

We also have the following corollaries to the theorem.

Corollary 1

If d is between $F_h/S_{t,h}$ and $F_{h-1}/S_{t,h-1}$ with $h \geqslant 3$, then $D_i = 3D_{i-2} - D_{i-4}$ for $i = 5, \ldots, h+1$.

Corollary 2

(A) If
$$F_{2k}/S_{t,2k} < d < F_{2k-1}/S_{t,2k-1} \text{ with } k \geqslant 1,$$
then for $i = 1, \ldots, k-1$,
$$\frac{F_{2k-2i}}{F_{2k-2i+2}} < \frac{D_{2i+1}}{D_{2i-1}} < \frac{F_{2k-2i-1}}{F_{2k-2i+1}}.$$

(B) If
$$F_{2k}/S_{t,2k} < d < F_{2k+1}/S_{t,2k+1} \text{ with } k \geqslant 1,$$
then for $i = 1, \ldots, k-1$,
$$\frac{F_{2k-2i}}{F_{2k-2i+2}} < \frac{D_{2i+2}}{D_{2i}} < \frac{F_{2k-2i-1}}{F_{2k-2i+1}}.$$

Corollary 3

(A) If
$$F_{2k}/S_{t,2k} < d < F_{2k-1}/S_{t,2k-1} \text{ with } k \geqslant 1,$$
then for $i = 1, \ldots, k-1$,
$$1/3 < D_{2i+1}/D_{2i-1} < 1/2.$$

(B) If
$$F_{2k}/S_{t,2k} < d < F_{2k+1}/S_{t,2k+1} \text{ with } k \geqslant 1,$$
then for $i = 1, \ldots, k-1$,
$$1/3 < D_{2i+2}/D_{2i} < 1/2.$$

Note that Corollary 3 is a weaker version of Corollary 2. With the same hypothesis as Corollary 2, it gives a cruder estimate of the range of permitted values of D_{h+1}/D_{h-1}. Corollary 3 describes a property of Fibonacci phyllotaxis that was described by Wright (1873, p. 399) when he said that in Fibonacci phyllotaxis "each leaf of the cycle is so placed over the space between older leaves nearest in direction to it as always to fall near the middle, and never beyond the middle third of the space ...". That it is a characteristic property of Fibonacci phyllotaxis can be seen in the following way: Suppose that there is a leaf distribution with constant divergence d such that $1/(t+1) < d < 1/t$, and that $1/3 < D_{h+1}/D_{h-1} < 1/2$ for $h = 2, \ldots, k$.

Then in particular we have $1/3 < D_{2i+1}/D_{2i-1} < 1/2$ for $i = 1, \ldots, j$, where $j = h/2$ if h is even, and $j = (h-1)/2$ if h is odd. It can be shown that under these conditions $F_{2j+2}/S_{t,\,2j+2} < d < F_{2j+1}/S_{t,\,2j+1}$. The proof is easily carried out by induction on j. This property of Fibonacci phyllotaxis plays a key role in Richards' (1948, pp. 225–6) attempt to explain the genesis of Fibonacci phyllotaxis. The reason why this explanation is inadequate is discussed in section 15.

In Corollary 2, if we let k increase to infinity, then

$$\lim_{k \to \infty} d = (t + \tau^{-1})^{-1},$$

and

$$\lim_{k \to \infty} D_{i+1}/D_{i-1} = \tau^{-2}$$

for all $i \geqslant 2$. $D_2 = 1 - td$. Consequently, when $d = (t + \tau^{-1})^{-1}$,

$$D_2 = 1 - t(t + \tau^{-1})^{-1} = \tau^{-1}(t + \tau^{-1})^{-1}.$$

Thus we have Corollary 4.

Corollary 4

When $d = (t + \tau^{-1})^{-1}$, $n_i = S_{t,\,i-1}$ for all $i \geqslant 1$.

Corollary 5

When $d = (t + \tau^{-1})^{-1}$, $D_i = (\tau^{-1})^{i-1}(t + \tau^{-1})^{-1}$ for $i \geqslant 1$.

Corollary 6

When $d = (2 + \tau^{-1})^{-1} = \tau^{-2}$, $D_i = \tau^{-i-1}$ for $i \geqslant 1$.

Using this last formula, we derive in Appendix C the following formula for the value of r when the opposed parastichy pair (n_{k-1}, n_k) or (n_k, n_{k-1}), i.e. (F_k, F_{k+1}) or (F_{k+1}, F_k), is conspicuous and orthogonal:

$$r = \tau^{-k-\frac{1}{2}}(F_k F_{k+1})^{-\frac{1}{2}}. \tag{20}$$

This formula, derived by using a cylindrical representation of a leaf distribution, is equivalent to the formula derived by Richards using a centric representation (see Appendix C). As this formula shows, the higher the parastichy numbers of the conspicuous opposed parastichy pairs, the lower the values of r. Therefore any model of phyllotaxis that would account for the observed fact that the genesis of the phyllotaxis of a plant proceeds from lower phyllotaxis (parastichy numbers) to higher phyllotaxis, must assume that r decreases for a while.

I. ADLER

9. Maximum Relative Leaf Diameter for Variable Divergence

In section 8 the maximum value of δ was identified, when d and r are fixed, as the distance from leaf 0 to its nearest neighbor. If d is allowed to vary, this distance will vary with it. Conversely, if this distance is allowed to vary, d must also vary to be compatible with it. Let us examine the consequences if δ grows, tending to be as large as possible.

Suppose the neighbor nearest to leaf 0 is the point of close return n_i. There are three possibilities to consider: (1) n_i is to the right of l_0. In this case, as δ grows, the center of leaf n_i is pushed to the right with respect to l_0, and this entails an increase in the divergence d; (2) n_i is to the left of l_0. In this case, as δ grows, the center of leaf n_i is pushed to the left with respect to l_0, and this entails a decrease in the divergence d; (3) n_i is on l_0. If r is fixed, δ could grow by pushing the center of n_i either to the right or to the left, so that growth of δ is compatible with either an increase or a decrease of d. However, we may exclude this ambiguous case from consideration, because it is highly improbable. For n_i to be on l_0, it would be necessary for d to be a rational number, and the set of rational numbers in any interval on the real number line is a set of measure zero. Moreover, in a growing plant, in which d varies with time, even if d were to have a rational value, it would soon change. It can be concluded then that growth of δ would compel d to *either* increase *or* decrease.

It is obvious that δ has an upper bound. In fact, the girth of the stem, which is equal to 1, is such an upper bound. Consequently, as d varies, there is a value of d at which δ is maximized. It is uniquely determined as the first value beyond which the distance from leaf 0 to its nearest neighbor would start decreasing. Thus we arrive at the mathematical assumption that corresponds to the biological assumption of the presence of contact pressure:

THE MAXIMIN PRINCIPLE

If contact pressure begins at $T = T_c$, then for $T \geqslant T_c$ and associated rise $r(T)$, d has the uniquely determined value at which the minimum distance from leaf zero to its nearest neighbor is maximized.

We can also derive a simple criterion for determining this value of d for a given value or r that is small enough. Suppose that n_i is the point of close return that is nearest to leaf 0. The fact that n_i is a point of close return implies that $n_i =$ leaf number q_i and d is between p_{i-1}/q_{i-1} and p_i/q_i, where p_{i-1}/q_{i-1} and p_i/q_i are the $(i-1)$th and ith principal convergents respectively of the simple continued fraction expansion for d (Coxeter, 1972). The fact that n_i is the leaf that is closest to leaf 0 implies that for some d between p_{i-1}/q_{i-1} and p_i/q_i dist $(0, n_i) \leqslant$ dist $(0, n_{i-1})$.

From the properties of continued fractions, we have

$$\text{dist}\,(l_0, n_i) = |q_i d - p_i|, \quad \text{dist}\,(l_0, n_{i-1}) = |q_{i-1}d - p_{i-1}|,$$

and

$$|p_i q_{i-1} - q_i p_{i-1}| = 1.$$

Let

$$f = \text{dist}^2\,(0, n_i) = (q_i d - p_i)^2 + q_i^2 r^2,$$
$$g = \text{dist}^2\,(0, n_{i-1}) = (q_{i-1}d - p_{i-1})^2 + q_{i-1}^2 r^2.$$

The graph of f as a function of d is a parabola with vertex at $(p_i/q_i, q_i^2 r^2)$. The graph of g is a parabola with vertex at $(p_{i-1}/q_{i-1}, q_{i-1}^2 r^2)$. The graphs cross in the interval between p_{i-1}/q_{i-1} and p_i/q_i provided that

$$f(p_i/q_i) < g(p_i/q_i).$$

This leads to the condition

$$r < 1/(q_i(q_i^2 - q_{i-1}^2)^{\frac{1}{2}}). \tag{21}$$

If r satisfies this condition, then dist $(0, n_i)$ is maximized at the value of d for which $f = g$. The condition $f = g, r \geqslant 0$ defines the following semicircle in the (d, r) plane:

$$(d - (p_i q_i - p_{i-1}q_{i-1})/(q_i^2 - q_{i-1}^2))^2 + r^2 = 1/(q_i^2 - q_{i-1}^2), \, r \geqslant 0. \tag{22}$$

We shall call it the (n_i, n_{i-1}) semicircle. Then we have this criterion for maximization of δ: If n_i is the point of close return that is nearest to leaf 0, and if r satisfies (21), then the point (d, r) lies on the (n_i, n_{i-1}) semicircle (The Maximin Criterion). In section 11 this criterion is employed repeatedly to identify conditions that suffice to produce normal phyllotaxis.

10. Contact Pressure and Phyllotaxis Biography

If a leaf distribution is a regular cylindrical lattice, its phyllotaxis at any instant T is completely determined by the values of d and r, and therefore may be represented by a point in a two-dimensional phase space, the (d, r) plane. The phyllotaxis biography of the leaf distribution is then represented by a path in the (d, r) plane with the parametric equations $d = d(T), r = r(T)$, where $d(T)$ and $r(T)$ are functions of the time. In the absence of contact pressure, these functions may be independent of each other. In the presence of contact pressure, they are not independent, because, for any given r, the value of d is determined by the Maximin Principle. The parametric equations then have this form: $d = d(r(T)), r = r(T)$. The nature of the path then depends on the following: (1) the initial values of (d, r) for $T < T_c$, where T_c is the time when contact pressure begins; (2) the time T_c when contact pressure begins; (3) the nature of the function $r(T)$. For any given plant, (1), (2) and (3) can be determined by observation. Then, assuming that the leaf distribution of the plant is a regular cylindrical lattice, the value of d that would be determined by contact pressure at each instant $T \geqslant T_c$ could be calculated by computer by making use of the Maximin Principle. Comparing the calculated values of d with the observed values of d at these times would then serve as a test of whether or not the observed changes in the phyllotaxis of the plant can be accounted for by contact pressure. However, the leaf distribution of a plant is not a regular lattice. Successive divergences d_n are not all equal. For the test to be applicable, it must first be modified to take this fact into account. A valid modification of this kind is developed in section 13. Anticipating the results of this section for the moment, we conclude that the mathematical model constructed here for a leaf distribution with contact pressure removes from the realm of speculation the question of whether or not contact pressure, where it exists, can account for the observed changes in phyllotaxis. This is now a question that can be settled by observation.

11. Conditions Under Which Contact Pressure Will Produce Normal Phyllotaxis

By definition, a divergence is less than $1/2$. The unit fractions $1/t$ divide the interval between 0 and $1/2$ into subintervals of the form $[1/(t+1), 1/t]$ in which the endpoints $1/(t+1)$ and $1/t$ are included, and each of them except $1/2$ belongs to two adjacent intervals. In order to eliminate this possible source of ambiguity, the unit fractions $1/t$ shall not be considered. (At the end of this section we shall explain why this exclusion is justified.) With the unit fractions excluded, every divergence belongs to a unique open interval of the form $[1/(t+1), 1/t]$. In this section, assuming that the initial divergence is in the interval $[1/(t+1), 1/t]$, conditions are identified under which contact pressure compels the divergence to converge toward the value $(t+\tau^{-1})^{-1}$, so that normal phyllotaxis is produced.

We begin with the assumption that $T_c < t+1$, i.e. that δ is maximized before leaf $t+1$ emerges. It is also assumed that r is a monotonic decreasing function of T. In section 9 we found the Maximin Criterion for maximization of δ: maximization of δ, when the point of close return n_i is the leaf nearest to leaf 0, requires that the point (d, r) in the (d, r) plane be on the (n_i, n_{i-1}) semicircle, *provided that r is small enough*. As r decreases, different points of close return n_i take on in succession the role of "leaf nearest to leaf 0." By repeated use of the Maximin Criterion a class of functions $r(T)$ is identified which assure that the path of the point (d, r) will be a zig-zag path made of arcs of successive (n_i, n_{i-1}) semicircles with the property that the projection of these arcs on the d-axis is a nest of intervals converging on the point $(t+\tau^{-1})^{-1}$.

The principal results are as follows. The *normal phyllotaxis path* (for given t) is defined as the succession of arcs of the $(S_{t,n}, S_{t,n-1})$ semicircles joining the sequence of points $P_n = (d_n, r_n)$ where the $(S_{t,n-1}, S_{t,n-2})$ semicircle and the $(S_{t,n}, S_{t,n-1})$ semicircle meet. It can be shown that

$$d_n = \frac{4F_{n-1}S_{t,n-1}+(2t-1)(-1)^{n-1}}{2[2S_{t,n-1}^2+(-1)^{n-1}(t^2-t-1)]},$$

$$r_n = \frac{(3)^{\frac{1}{2}}}{2[2S_{t,n-1}^2+(-1)^{n-1}(t^2-t-1)]}. \tag{23}$$

Defining s_n as the value of r such that when $r < s_n$, leaf $S_{t,n}$ is the next point of close return after leaf $S_{t,n-1}$,

$$s_n = \frac{(3)^{\frac{1}{2}}}{S_{t,n+1}(S_{t,n-3}S_{t,n})^{\frac{1}{2}}}. \tag{24}$$

We then establish

THEOREM 12

If (1) *r is decreasing,* (2) *the relative leaf diameter is maximized before $T = t+1$, and* (3) *for every integer $n > 1$ leaf $S_{t,n}$ emerges when $r_n \leqslant r < s_n$, where r_n is defined by* (23) *and s_n is defined by* (24), *then, from the time leaf t is fully grown, the point (d, r) is always on the normal Fibonacci path, and d approaches $(t+\tau^{-1})^{-1}$ as a limit as $n \to \infty$.*

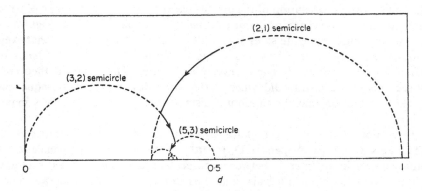

FIG. 4. The normal phyllotaxis path for $t = 2$.

Full details of the proof for the case $t = 2$ (Theorem 9) are given in Appendix D. The normal phyllotaxis path for $t = 2$ is shown in Fig. 4. From Theorem 9 we then derive the following main theorem.

THEOREM 10

If the relative leaf diameter δ of the leaf primordia is maximized at time $T_c < 3$, and if the rise $r(T)$ at time T satisfies

$$1 \cdot 14 \leqslant T^2 r(T) < 1 \cdot 79 \qquad (25)$$

for $3 \leqslant T \leqslant F_N$, then (d, r) is on the normal phyllotaxis path throughout the interval of time $T_c \leqslant T \leqslant F_N$, and the plant develops higher and higher normal phyllotaxis at least through (F_{N-1}, F_{N-2}) phyllotaxis.

The proof is given in Appendix D.

In sections 8–11 it was assumed that leaves are circular regions as it was a convenient way of picturing the size of a leaf. We now drop this assumption and show that the theorems of these sections (and the corresponding appendices) hold even when we take into account the fact that the shape of each leaf is changed under contact pressure. Whatever the shape of the leaves, it is assumed that they will grow in size until the two nearest ones come into contact. After that, with continued growth of the leaves, each will exert a force on the other. This force will deform the boundary of each leaf at the point of contact, tending to flatten it. At the same time, however, it will tend to push the leaves apart until the distance between their centers becomes as great as possible. Then, with further growth, each leaf will grow closer to its second nearest neighbor, until they make contact and their boundaries are flattened at this point of contact too. In the final stage of growing toward each other, the leaves would grow into all remaining spaces between leaves until the leaves form a tesselation of the cylindrical surface. The final shape of each leaf, jammed up tight against its nearest neighbors, will be a parallelogram or a hexagon whose opposite sides are parallel. The important point is that even though the shape of the leaves is changed, *the pressure of growth tends to maximize the minimum distance between leaf centers.* Therefore the theorems of sections 8–11 are still valid, since all they do is develop some consequences of the assumption that the minimum distance between leaf centers is maximized.

One more minor point remains to be clarified. In the hypotheses of Theorems 8 to 11 of Appendix D, in order to locate each divergence d unambiguously in an interval between consecutive unit fractions $1/(t+1)$ and $1/t$, we used the open intervals, and thus excluded from consideration the unit fractions themselves. This can be justified by the fact that it is highly improbable that d will be exactly a unit fraction. (See the argument in section 9 explaining why it is improbable that d is rational.)

12. Conditions Under Which Contact Pressure Will Produce Anomalous Phyllotaxis

Theorem 9 (section 11) established that when $t = 2$, if the relative leaf diameter is maximized when $T < 3$, and $r(T)$ belongs to a particular class of functions, then the point (d, r) moves on the normal phyllotaxis path. A similar analysis can be made to determine how the phyllotaxis of a plant may change with time if the maximization of the relative leaf diameter is delayed to some time $T \geqslant 3$. Under this assumption there are some conditions that will also produce normal phyllotaxis, with (d, r) moving toward the normal phyllotaxis path and then, after reaching it, moving along it. But there are also other conditions capable of producing a different kind of phyllotaxis. To demonstrate the latter, we shall identify some conditions that produce anomalous phyllotaxis. We draw the $(2, 1)$, $(3, 2)$, $(3, 1)$, $(5, 1)$, $(5, 2)$ and $(5, 3)$ semicircles in the (d, r) plane. These divide the plane into regions. Two of these regions are numbered I and II respectively in Fig. 5. R is defined as

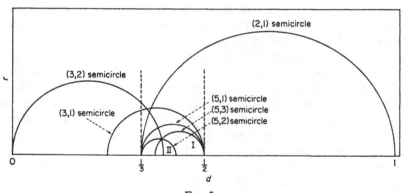

FIG. 5.

the rectangle whose base is the interval $2/5 < d < 1/2$ on the d-axis, and whose height is $(21)^{\frac{1}{2}}/105$. The following results are established. Contact pressure will produce anomalous phyllotaxis first as $(2, 5)$ phyllotaxis, and then as $(7, 5)$ phyllotaxis, under the following conditions: (1) $5 \leqslant T_c < 6$; (2) at time T_c, (d, r) is in the region $R \cap (I \cup II)$; (3) leaf 7 emerges and is fully grown when $(3)^{\frac{1}{2}}/78 \leqslant r < (7)^{\frac{1}{2}}/84$, and the next close return candidate after leaf 7 emerges when $r < (3)^{\frac{1}{2}}/78$. Full details are given in Appendix E.

The analysis can be carried further step by step to identify conditions after $T = 7$ that would compel the phyllotaxis to climb to higher values (7, 12), (7, 19), ... of anomalous phyllotaxis. These conditions would provide us with a theorem for anomalous phyllotaxis that is analogous to Theorem 9 for normal phyllotaxis.

It is an interesting fact that the ideal angles for both normal phyllotaxis and anomalous phyllotaxis can be given by a single formula. For the normal sequence 1, t, $t+1$, $2t+1$, ... the ideal angle is $(t+\tau^{-1})^{-1}$, $t = 2, 3, 4, \ldots$. For the anomalous sequence t, $2t+1$, $3t+1$, $5t+2$, ... the ideal angle can be shown to be $-[-(t+1)+\tau^{-1}]^{-1}$, $t = 2, 3, 4, \ldots$. Therefore both types of sequence are covered by the same formula,

$$\text{ideal angle} = |(s+\tau^{-1})^{-1}|, \tag{26}$$

with $t = s \geqslant 2$ for the normal sequence, and $-(t+1) = s \leqslant -3$ for the anomalous sequence. (I am indebted to Arthur Winfree for suggesting that the meaning of negative values of s in (26) be investigated.)

13. Tests of the Contact Pressure Model

In sections 8–12 a mathematical model of a leaf distribution with contact pressure has been developed. The model assumes that successive divergences d_n are equal. However, on the growing tip of a stem, successive values of d_n are not all equal, although, with increasing n, they tend toward approximate equality. It is necessary to eliminate this discrepancy between the model and growing plants before a practical test of the model can be devised. The discrepancy can be eliminated if we can determine from the unequal successive values of d_n in a plant a single number that would correspond to the d of the model. We can do this by identifying d at time $T = j$ with the average of $d_1, \ldots, d_j$. Then, given a plant in which contact pressure exists, it can be determined by the following sequence of steps whether contact pressure can account for the observed divergences in the plant:

(1) Determine by observation the time T_c when contact pressure begins. Let k be the largest integer in T_c.

(2) Measure $d_1, \ldots, d_k$ at time $T = T_c$, and calculate their average.

(3) Define $d(T_c) = $ the average of $d_1, \ldots, d_k$.

(4) Measure $r(T)$ (the value of r at time T), for $T \geqslant T_c$, at intervals of time equal, say, to 0·1 plastochrone.

(5) Define $d(T)$ for $T > T_c$ inductively as follows: Let $d = d(T-0\cdot1)$; let $r = r(T)$. Using these values of d and r and the number of leaves present, determine which leaf n is nearest to leaf 0. Choose some increment ε for d,

say $\varepsilon = 0.001$. Calculate dist $(0, n)$ for $d = d(T-0.1)$. Determine whether dist $(0, n)$ is increased when d is increased or decreased by ε. (In general, at most one of these cases will occur.) Suppose, for example, dist $(0, n)$ is increased when d is increased. Then calculate dist $(0, n)$, using for d the successive values $d+\varepsilon$, $d+2\varepsilon$, etc., until either dist $(0, n)$ stops increasing or n ceases to be the leaf that is nearest to leaf 0. Let $d+m\varepsilon$ be the value of d that yields the highest value of dist $(0, n)$ while n is the leaf nearest to leaf 0. Then define $d(T) = d(T-0.1)+m\varepsilon$. A computer program for calculating $d(T)$ can easily be written.

(6) If h is the largest integer in T, measure $d_1, \ldots, d_h$ at time T, and calculate their average $\bar{d}(T)$.

(7) Compare $d(T)$ with $\bar{d}(T)$. Agreement of $d(T)$ with $\bar{d}(T)$ for $T > T_c$ would show that contact pressure suffices to account for the changes in the plant's phyllotaxis after $T = T_c$.

The model also makes the following additional testable predictions when the growth rates of δ and r satisfy certain special conditions. (In these predictions, interpret d at time T to mean the average divergence at time T.)

If the growth rates of r and δ are such that δ is maximized before $T = 3$, and $1.14 \leqslant T^2 r(T) < 1.79$ for $3 \leqslant T \leqslant F_N$ for some $N \geqslant 4$;

(1) The plant develops higher and higher normal phyllotaxis up to (F_{N-1}, F_{N-2}) phyllotaxis.

(2) The divergence d is a function of r, and when the phyllotaxis is (F_n, F_{n-1}) with $4 \leqslant n < N$, the points (d, r) lie on the (F_n, F_{n-1}) semicircle defined by (D14) in Appendix D.

(3) Left and right conspicuous parastichies at each stage of development after the maximization of δ have equal steps (distances between adjacent leaf centers on the parastichies).

(4) While the phyllotaxis is (F_n, F_{n-1}), with $4 \leqslant n < N$, δ^2 is a linear function of d given by (D19) in Appendix D.

(5) The divergence d approaches the ideal angle by alternately decreasing and increasing.

(6) Whereas in an arbitrary cylindrical lattice in which (F_n, F_{n-1}) is a conspicuous opposed parastichy pair the divergence may have any value between F_{n-3}/F_{n-1} and F_{n-2}/F_n, the model, with the growth rates assumed above, permits the divergence to be only in the narrower range between

$$\frac{4F_{n-2}^2+(-1)^n}{2(2F_{n-1}^2+(-1)^{n-1})} \quad \text{and} \quad \frac{4F_{n-1}^2+(-1)^{n+1}}{2(2F_n^2+(-1)^n)},$$

for $4 \leqslant n < N$. For example, an arbitrary cylindrical lattice with $(5, 3)$ phyllotaxis permits $0.3333 < d < 0.4000$, while the model (with the assumed growth rates) permits only $0.3776 < d < 0.3947$.

If δ is maximized only after $T = 5$ and when (d, r) is in the region $R \cap (I \cup II)$ defined in section 12:

(7) The plant develops (5, 2) phyllotaxis, an example of anomalous phyllotaxis.

14. More than One Fundamental Spiral

Assume a leaf distribution represented by a cylindrical lattice on a cylindrical surface with girth 1. Let 0 be an arbitrary leaf of the lattice. The number of leaves at the same level as 0 is the number of fundamental spirals of the leaf distribution. Let 0_1 be the nearest leaf to 0 on the same level as 0 to the right of 0 as seen from outside the cylindrical surface. (If there is only one fundamental spiral, 0_1 will coincide with 0.) Develop in a plane the strip of the cylindrical surface generated by the element through 0 when it moves to the right from 0 to 0_1. If g is the number of fundamental spirals, the width of this strip is $1/g$. The theorems established in this paper are really theorems about such a plane lattice strip whose width is assumed to be 1. They can be converted to theorems about a leaf distribution with g fundamental spirals with $g \geqslant 1$ by merely assuming that the width of the strip is $1/g$. For example, Theorem 4 generalizes in this way to the following theorem for a leaf distribution with g fundamental spirals: The opposed parastichy pair $(gS_{t,h+1}, gS_{t,h})$ is visible if and only if gd equals $F_h/S_{t,h}$ or $F_{h+1}/S_{t,h+1}$ or lies between them. This reduces to Theorem 4 when $g = 1$.

15. What a Field Theory Can and Cannot Do

Schoute (1913) proposed an inhibitor field theory for the limited purpose of explaining why a leaf distribution tends to approximate a regular lattice. He assumed that contact pressure would explain why the divergence of the lattice tends to be approximately $(t + \tau^{-1})^{-1}$ so that the phyllotaxis becomes normal Fibonacci phyllotaxis. Richards (1948, 1951) cast the field theory for a wider role. He claimed that a field theory alone, without benefit of contact pressure, sufficed to explain both phenomena, the approximate regularity of the lattice, and the widespread occurrence of normal Fibonacci phyllotaxis. If Richards were right, then everything said in sections 8–13 would be superfluous and irrelevant. It is therefore necessary to deal with this claim.

The Richards field theory postulates the secretion of an inhibitor by the stem apex and by each existing leaf primordium. Diffusion of the inhibitor from these sources results in a variable concentration of the inhibitor at each point of the meristem, varying with both position and time. Richards did not attempt to write any field equations that would describe explicitly how the concentration of the inhibitor varies with position and time. However, there are implicit in his discussions of a field theory, certain qualitative assumptions that characterize the behavior of the inhibitor. When stated in terms of a cylindrical representation of a leaf distribution, the Richards assumptions, which we shall call assumptions R, take the form:

(1) The level at which a new primordium will arise is determined chiefly by the inhibitor secreted by the stem apex. The concentration of this inhibitor decreases as the distance from the stem apex increases. The primordium will arise at that level where the concentration has attenuated below a critical level.

(2) At that level, the primordium arises where the sum of the concentrations of the inhibitors secreted by the older leaf primordia is a minimum.

(3) The concentration of the inhibitor from a single leaf primordium is a decreasing function of the distance from the center of the primordium.

Richards argued that it follows from these assumptions that (a) with the level of the new primordium determined chiefly by the stem apex, its position at that level is determined principally by the two primordia that are its nearest neighbors; and (b) the primordium is formed "tangentially somewhat nearer the older one" of these two determining primordia. He then concluded that it follows from (b) that the successive divergences of the new leaf primordia converge toward the ideal angle of normal Fibonacci phyllotaxis. Let us call this conclusion "conclusion NF". The essence of the Richards theory then is the assertion that assumptions R imply conclusion NF. This assertion, since it is a statement of implication, is a *mathematical assertion*. Hence proving it or disproving it is a mathematical problem, not a biological problem. We shall disprove it by producing a counter-example.

At any point P on the level at which a new primordium emerges, let $\phi(P, T)$ be the concentration of the inhibitor due to a single primordium, where T is the age of that primordium in plastochrones. Let us assume

(I) ϕ can be factored into a position-dependent factor $f(P)$ and a time-dependent factor $h(T)$, where $f(P)$, $h(T) \geq 0$;

(II) $h(T) = 0$ for $T \geq 4$, and $h(1) > 0$.

Let x and y be the horizontal and vertical co-ordinates respectively of P in the plane development of the cylindrical surface using the center of the inhibiting primordium as origin, and write $f(x, y)$ for $f(P)$. Consider the situation when leaves $0, 1, \ldots, n-1$ are present and leaf n is about to emerge at the level whose height above leaf $n-1$ is r. If P is any point on that level, let $F(P, r)$ be the total concentration at P of the inhibitors released by the primordia $0, 1, \ldots, n-1$. Then

$$F(P, r) = h(1)f(x, r) + h(2)f(x + d_{n-1}, 2r) + h(3)f(x + d_{n-1} + d_{n-2}, 3r). \quad (27)$$

Then d_n, with $0 < d_n \leqslant 1/2$, is determined as the value of x for which

$$\frac{\partial}{\partial x} F(P, r) = 0,$$

provided that such a value exists and is unique, and that

$$\frac{\partial^2}{\partial x^2} F(P, r) > 0$$

for that value of x. Then d_n is determined by the recurrence relation

$$f_x(d_n, r) + af_x(d_n + d_{n-1}, 2r) + bf_x(d_n + d_{n-1} + d_{n-2}, 3r) = 0, \qquad (28)$$

where $a = (h(2))/(h(1))$, $b = (h(3))/(h(1))$.

Considerations of continuity suggest that we assume

(III) For small values of r, there exist values of a and b for which d_n is given approximately by the simpler equation obtained by taking $r = 0$. Let

$$g(x) = -f_x(x, 0). \qquad (29)$$

Then d_n is defined by

$$g(d_n) + ag(d_n + d_{n-1}) + bg(d_n + d_{n-1} + d_{n-2}) = 0. \qquad (30)$$

Assume that

(IV) $f_x(x, 0) < 0$ for $0 < x < 1/2$, and $f_x(0, 0) = -\infty$, (so that $g(x) > 0$ for $0 < x < 1/2$, and $g(0) = \infty$).

(V) $g(x)$ is a monotonic decreasing continuous function of x, and $g(1-x) = -g(x)$, for $0 < x < 1$.

(VI) $h(3) \leqslant h(2)$, (so that $b \leqslant a$).

(VII) Among the values of a and b satisfying (III), there exist some for which (30) determines a unique sequence (d_n) with the property $0 < d_n \leqslant 1/2$, for $n = 1, 2, 3, \ldots$, and $d_n + d_{n-1} + d_{n-2} < 1$ for $n \geqslant 3$.

Designating assumptions (I) to (VII) collectively as assumption 3′, the set of functions $f(x, y)$ which, together with appropriate functions $h(T)$, satisfy 3′ is not empty, e.g., $f(x, y) = 1/s + 1/s'$ is such a function, where $s = (x^2 + y^2)^{\frac{1}{2}}$ and $s' = [(1-x)^2 + y^2]^{\frac{1}{2}}$.

Designating as assumptions A the set of assumptions 1, 2 and 3′, since 3′ implies 3, assumptions A imply assumptions R. If Richards were correct in his assertion that assumptions R imply conclusion NF, then it would follow that assumptions A imply conclusion NF. However, it will be shown that conclusion NF is false for functions satisfying assumptions A. In fact, it is shown that under assumptions A successive divergences d_n converge to a limit d which may be *any number* in the range $0 \cdot 2 < d < 0 \cdot 5$, the value of d depending on the values of a and b. This clearly refutes Richards' assertion that d can only have the values in this range that are of the form $(t + \tau^{-1})^{-1}$, namely $(2 + \tau^{-1})^{-1}$, $(3 + \tau^{-1})^{-1}$, and $(4 + \tau^{-1})^{-1}$.

There are three cases to be considered:

Case I (one plastochrone case): $a = b = 0$.
Case II (two plastochrone case): $a > 0, b = 0$.
Case III (three plastochrone case): $a > 0, b > 0$.

Case I (one plastochrone case). $a = b = 0$. Equation (30) reduces to $g(d_n) = 0$, and has the unique root $d_n = 1/2$.

Case II (two plastochrone case). $a > 0, b = 0$. At the time that leaf number 1 emerges, leaf number 0 is one plastochrone old, and there are no leaves present whose age is two plastochrones or more. Consequently, the one plastochrone case applies, and $d_1 = 0.5$. For $n > 1$ we have

$$g(d_n) + ag(d_n + d_{n-1}) = 0. \tag{31}$$

We then have the following theorem.

THEOREM A

If $d_1 = 0.5$, and d_n for $n > 1$ is defined by (31), then the sequence (d_n) is well-defined and converges to a limit d with $0.25 < d < 0.5$. d and a are related by the condition

$$a = -\frac{g(d)}{g(2d)}. \tag{32}$$

The proof is given in Appendix F.

Case III (three plastochrone case). $0 < b \leqslant a$. From the one plastochrone case, $d_1 = 0.5$. d_2 is determined by equation (31) of the two plastochrone case; and for $n > 2$, d_n is determined by equation (30).

We then have the following theorem.

THEOREM B

If $d_1 = 0.5$, d_2 is defined by equation (31), and d_n for $n > 2$ is defined by equation (30), then the sequence (d_n) is well-defined and converges to a limit d with $0.2 < d < 1/3$. Values of a and b that determine any given value of d in this range satisfy the linear equation

$$g(d) + ag(2d) + bg(3d) = 0. \tag{33}$$

The proof is given in Appendix F.

Values of d determined by computer if $f(x, y) = 1/s + 1/s'$ are shown in Tables 6, 7 and 8 in Appendix F. These tables show that, when r is small, for appropriate values of a and b, the error in the value of d is negligible if we assume

32   I. ADLER

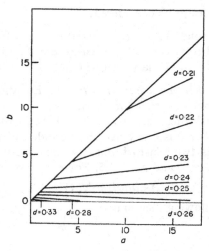

FIG. 6. The three-plastochrone case (calculated with $r = 0$) when $f(x, y) = 1/s + 1/s'$.

$r = 0$. Taking $r = 0$, the relationship between d and (a, b) is summed up in Fig. 6. Any function $f(x, y)$ satisfying sssumption 3' will produce a similar graph.

It is important to note that we are not proposing $1/s + 1/s'$ as the appropriate form for the function $f(x, y)$ that gives the position-dependent factor of the inhibitor function ϕ. In fact, if ϕ is determined by a diffusion process, it is likely that $f(x, y)$ does not have this form. However, this does not preclude the use of $1/s + 1/s'$ as part of a counter-example that proves that assumptions R do not imply conclusion NF.

While proving that assumptions R do not suffice to explain the genesis of normal phyllotaxis, we have derived another incidental benefit. It has been shown that there exist functions satisfying assumptions R which require that d_n converge to a limit d. It follows, then, that a field theory of the type envisaged by Richards may be able to explain why a leaf distribution tends to approximate a regular lattice, even though it cannot provide the whole explanation for the genesis of normal phyllotaxis. This supports the orginal conjecture of Schoute.

It is relevant to ask what might have led Richards to think that assumptions R imply conclusion NF. As noted earlier in this section, Richards believed that assumptions R imply his intermediate conclusion (b), and he asserted that (b) implies conclusion NF. This assertion made on page 223 of Richards (1948) appears to be based on a misunderstanding of the property noted by Wright (1873) as a characteristic property of Fibonacci phyllotaxis and which we have formulated precisely in Corollaries 2 and 3 of Theorem 7 (and in a converse of Corollary 3) in section 8.

(a) Although the property $1/3 < D_n/D_{n-2} < 1/2$ which is mentioned in Corollary 3 of Theorem 7 is a characteristic property of normal Fibonacci phyllotaxis, it is not a sufficient condition for producing it. It becomes sufficient only when combined with the assumption that successive divergences are equal, e.g. if $1/3 < d < 1/2$, we have for the first two points of close return $n_1 = 1$, $n_2 = 2$, and $n'_3 = 3$ is the next close return candidate after leaf 2. If $d_1 = d_2 = d_3 = d$, then $D_3 = 3d-1$. Then the condition $1/3 < D_3/D_1 < 1/2$ yields $d/3 < 3d-1 < d/2$, which implies $3/8 < d < 2/5$. Using Richards' rule alone for inserting leaves 1, 2 and 3, without assuming $d_1 = d_2 = d_3 = d$, this will not suffice to produce an average divergence that is between 3/8 and 2/5. In fact, if we let $d_1 = 0.48$, $d_2 = 0.255$, and $d_3 = 0.495$, although we have $1/3 < 0.48 < 1/2$, $1/3 < 0.255/0.52 < 1/2$, and $1/3 < 0.23/0.48 < 1/2$, the average divergence $= (1/3)(d_1+d_2+d_3) = 0.411 > 2/5$.

(b) Richards either overlooked or did not know that the rule expressed in Corollary 3 of Theorem 7 was only a weaker formulation of a stronger property expressed in Corollary 2. For example, suppose we have successive leaves inserted so that $1/3 < D_1 < 1/2$, $1/3 < D_3/D_1 < 1/2$, and $1/3 < D_5/D_3 < 1/2$. Then it can be shown that if the divergences are equal to d that $8/21 < d < 5/13$. But it is also necessary then that $3/8 < D_3/D_1 < 2/5$. Thus making $1/3 < D_3/D_1 < 1/2$ and $1/3 < D_5/D_3 < 1/2$ is not sufficient to produce $8/21 < d < 5/13$, because the condition $1/3 < D_3/D_1 < 1/2$ has to be replaced by the stronger condition $3/8 < D_3/D_1 < 2/5$. In other words, to make the Richards rule work to produce a closer and closer approach of the divergence to the ideal angle, every time a gap is divided according to the "more than 1/3, less than 1/2" rule, it is necessary to go back and readjust the ratios of the consecutive gaps under it to compel them to be in narrower ranges than were permitted before. But the Richards rule does not provide any mechanism for readjusting the widths of the gaps between leaves *after* the leaves are formed.

16. Toward a Mathematical Model of Phyllotaxis

The mathematical model of contact pressure developed in sections 8–13 is not a complete mathematical model of phyllotaxis. It assumes without explanation that successive divergences tend toward approximate equality. But with this assumption, the model identifies conditions under which contact pressure suffices to produce normal Fibonacci phyllotaxis. On the other hand, a field theory of the Richards type is also not a complete mathematical model of phyllotaxis. It has been seen that it is not capable of explaining the development of normal Fibonacci phyllotaxis, but, with certain assumptions about the inhibitor field function, it is capable of explaining why successive divergences tend toward approximate equality. However, an appropriate field theory combined with the contact pressure model developed here could serve as a complete mathematical model of phyllotaxis, because each supplies what the other lacks.

But we do not yet have an appropriate field theory. A fully developed field theory can be obtained only when the inhibitor field function $\phi(P, T)$ has been identified, so that it will be possible to derive an equation corresponding to equation (30) of section 15 by which the value of d_n is determined by the values of $d_1, \ldots, d_{n-1}$.

There are three avenues of investigation that may converge toward the identification of the inhibitor field function ϕ.

(1) Experimental biochemical investigation of leaf primordia may determine whether or not there really is an inhibitor that is secreted by the primordia. If there is one, and it is identified, then ϕ can be determined by measuring how the concentration of the inhibitor varies with position and time.

(2) Mathematical investigation of the diffusion equation may identify the function ϕ. What seems to be required is to find whether there are solutions of the diffusion equation on a cylinder with properties (I) to (VII) listed in section 15.

(3) Clues to the identity of ϕ may also be obtained from empirical studies of plants such as *Dryopteris*, where there does not seem to be any contact pressure. In the absence of contact pressure, the successive values of d_n are presumably determined by a recurrence relation like equation (30) of section 15. The empirically determined values of $d_1, \ldots, d_n, \ldots$ provide clues to the nature of the recurrence relation, and the recurrence relation is a differential equation for the function ϕ.

If and when there is an adequate field theory specifying the form of the function ϕ, it will be necessary to refine the contact pressure model before they can be combined to form a complete mathematical model of phyllotaxis A refinement would be needed to take into account the fact that successive values of d_n derivable from ϕ need not be equal. A refinement that may be appropriate would be replacing the Maximin Principle of section 9 by this generalization of it.

Generalized Maximin Principle

If contact pressure begins at $T = T_c$, then for $T \geqslant T_c$ and associated rise $r(T)$, if n is the largest integer in T, $d_1, \ldots, d_n$ have those values (in the neighborhood of immediately preceding values) for which the sum of the squares of the distance from each leaf to its nearest neighbor is maximized.

When $d_1 = d_2 = \ldots = d_n = d$, this Generalized Maximin Principle reduces to the Maximin Principle of section 9.

A complete mathematical model of phyllotaxis of the type described above would be testable, because it would make specific predictions of how the phyllotaxis of the plant at any time T should be related to the observed rise $r(T)$.

17. The History of Phyllotaxis

The discovery of the principal facts of phyllotaxis, and attempts to explain these facts both have a long history. A brief outline of this history is given in Appendix G, together with some critical comments about earlier theories.

I am indebted to H. S. M. Coxeter who read the first draft of this paper and made many helpful suggestions, and whose paper (1972) inspired the idea that the key to the riddle of Fibonacci phyllotaxis may be in the role of the points of close return. I wish to thank Stephen L. Adler, Chandler Davis, Edward Flaccus, Leo Gross, Harry Ruderman, Harry Sitomer, and Robert H. Woodworth, who also read the first draft and made useful suggestions. I am grateful to Robin O. Gandy for letting me have a copy of the unpublished notes of Turing on phyllotaxis. I also wish to thank Elizabeth Cutter, C. R. Illingworth, Bernard Richards, Claude Wardlaw, and the library staffs of the Bennington Free Library, the British Museum (Natural History), and the New York Botanical Garden for help in locating and obtaining copies of early papers on phyllotaxis. I am also grateful to Peggy Adler for the figures.

REFERENCES

AIRY, H. (1873). *Proc. R. Soc.* **21,** 176.
BONNET, C. (1754). *Recherches sur l'Usage des Feuilles dans les Plantes,* p. 159. Goettingue et Leyde: Luzac.
BRAUN, A. (1831). *Nova Acta Acad. Caesar. Leop. Carol.* **15,** 197.
BRAUN, A. (1835). *Flora, Jena.* **18,** 145.
BRAVAIS, L. & A. (1837). *Annls Sci. nat. Botanique* (2) **7,** 42; **8,** 11.
CHURCH, A. H. (1904). *On the Relation of Phyllotaxis to Mechanical Laws.* London: Williams & Norgate.
CHURCH, A. H. (1920). *On the Interpretation of Phenomena of Phyllotaxis.* Oxford University Press.
COXETER, H. S. M. (1961). *Introduction to Geometry,* p. 160. New York: Wiley.
COXETER, H. S. M. (1972). *J. Algebra* **20,** 167.
DAVIES. P. A. (1939). *Am. J. Bot.* **26,** 67.
DE CANDOLLE, C. (1881). *Considerations sur l'Etude de la Phyllotaxie.* Geneva: Georg.
FREY–WYSSLING, A. (1954). *Nature, Lond.* **173,** 596.
FUJITA, J. (1937). *Bot. Mag., Tokyo* **51,** 480.
HOGGATT, V. E. (1969). *Fibonacci and Lucas Numbers.* Boston: Houghton Mifflin.
LE VEQUE, W. J. (1956). *Topics in Number Theory,* p. 155. Reading: Addison–Wesley.
LUDWIG, F. (1896). *Bot. Zbl.* **68,** 7.
MACCURDY, E. (1955). *The Notebooks of Leonardo da Vinci,* p. 301. New York: Braziller.

36

PLINY. (1856). *Natural History*, Vol. 5, p. 227. (tr. by J. Bostock & H. T. Riley). London: Bohn.
RICHARDS, F. J. (1948). *Symp. Soc. exp. Biol.* **2**, 217.
RICHARDS, F. J. (1951). *Phil. Trans. R. Soc. B* **235**, 509.
SACHS, J. (1882). *Text-Book of Botany*, p. 201. London: Oxford University Press.
SACHS, J. (1887). *Lectures on the Physiology of Plants*. Oxford: Clarendon Press.
SACHS, J. (1906). *History of Botany* 1530–1860, p. 162. London: Oxford University Press.
SCHIMPER, K. F. (1830). *Geiger's Mag. für Pharm.* **29**, 1.
SCHOUTE, J. C. (1913). *Recl. Trav. bot. néerl.* **10**, 153.
SCHWENDENER, S. (1878). *Mechanische Theorie der Blattstellungen*. Leipzig: Engelmann.
SNOW, M. & R. (1962). *Phil. Trans. R. Soc. B* **244**, 483.
TAIT, P. G. (1872). *Proc. R. Soc. Edinb.* **7**, 391.
THEOPHRASTUS. (1916). *Enquiry into Plants*, p. 75. (tr. by A. Hort). New York: Putnam.
THOMPSON, D. W. (1942). *On Growth and Form*, Vol. 2, p. 912. 2nd edn. reprinted 1968. Cambridge University Press.
TURING, A. M. (unpublished). *Morphogen Theory of Phyllotaxis*. (Posthumous notes, R. O. Gandy, ed., to be published in *The Collected Works of A. M. Turing*. Amsterdam: North Holland Publ. Co.).
TURING, S. (1959). *Alan M. Turing, Mathematician and Scientist*, p. 141. Cambridge: Heffer.
VAN ITERSON, G. (1907). *Mathematische und Mikroskopische-Anatomische Studien der Blattstellungen*. Jena: Fischer.
WARDLAW, C. W. (1968). *Essays on Form In Plants*, p. 85. Manchester University Press.
WIESNER, J. (1875). *Flora, Jena* **58**, 113.
WIESNER, J. (1902). *Ber. dt. bot. Ges.* **20**, 84.
WIESNER, J. (1903). *Biol. Zbl.* **23**, 209.
WRIGHT, C. (1873). *Mem. Am. Acad. Arts Sci.* **9**, 379.

APPENDIX A

Some Fibonacci Identities

$$S_{t,h} = F_h t + F_{h-1}. \tag{A1}$$

$$S_{2,h} = F_{h+2}. \tag{A2}$$

$$3S_{t,h} - S_{t,h-2} = S_{t,h+2}. \tag{A3}$$

$$F_{n-k}F_{n+k} - F_n^2 = (-1)^{n+k+1}F_k^2. \tag{A4}$$

$$F_n F_{n-3} = F_{n-1}F_{n-2} + (-1)^n. \tag{A5}$$

$$F_{n-1}F_{n+1} - F_{n+2}F_{n-2} = 2(-1)^n. \tag{A6}$$

$$F_n F_{n+1}F_{n+3}F_{n+4} = F_{n+2}^4 - 1. \tag{A7}$$

$$F_{2n} = F_{n+1}^2 - F_{n-1}^2. \tag{A8}$$

$$F_{2n} = F_n L_n. \tag{A9}$$

$$5F_n^2 = L_n^2 - 4(-1)^n. \tag{A10}$$

$$F_n < \tau^n 5^{-\frac{1}{2}} + 1/2 < F_n + 1. \tag{A11}$$

APPENDIX B

Fundamental and Secondary Spirals in a Cylindrical Lattice

Each lattice point is called a "leaf".

(A) DEVELOPMENT IN A PLANE

Imagine the cylindrical surface slit along the element through 0 and unrolled onto a plane. Then the entire surface with all its leaves is contained in an infinite strip of width 1 closed only at the left edge. Use the left edge of the strip and the line containing the level of 0 as vertical and horizontal axes respectively for the plane. Repeat the strip indefinitely to the right and to the left to fill out the plane. Then we obtain a plane lattice in which no two leaves that are on adjacent levels are on the same vertical line. Every leaf b is repeated over and over again at unit intervals on the horizontal line through b, thus giving us in the plane infinitely many images of b. Denote by b_i the image of b whose directed distance from the b in the original strip is i. In particular, $b_0 = b$. In this plane development with repetitions, each of the fundamental spirals of the cylindrical surface becomes a straight line repeated infinitely many times.

(B) ELEMENTARY PROPERTIES

(1) If P and Q are leaves, then if there are leaves between P and Q, they divide line segment PQ into congruent segments.

(2) If P and Q are leaves and there is no leaf between P and Q, then if R is a point on the line PQ whose distance from P is an integral multiple of the distance PQ, then there is a leaf at R.

(3) If P, Q, R and S are vertices of a parallelogram, and there are leaves at P, Q and R, then there is a leaf at S.

(C) OPPOSED PARASTICHIES AND VISIBILITY

The word *parastichies* is used to refer to either parastichies or orthostichies. We begin with a series of definitions.

Definition

For a given opposed parastichy pair (m, n) and positive integer i, let the right parastichy through 0 and the left parastichy through 0_i intersect at P. Then triangle OPO_i is called an *opposed parastichy triangle with base i* belonging to the opposed parastichy pair (m, n). (Note that i is the length of the base 00_i of triangle OPO_i.) P is called the apex and OP and O_iP are called the legs of the opposed parastichy triangle. We shall use the abbreviation o.p. triangle for opposed parastichy triangle. A triangle OPO_i will be called simply a *parastichy triangle* (not necessarily opposed) if the intersecting parastichies OP and O_lP are not necesssarily opposed (right and left respectively).

Definition

An o.p. triangle with base 00_i belonging to the opposed parastichy pair (m, n) is called *matched* if there is a leaf at its apex and there are m steps on OP and n steps on O_lP.

Proposition 1

An opposed parastichy pair (m, n) is visible if and only if there is a matched o.p. triangle with base 1 that belongs to it.

Definition

A matched o.p. triangle with base 1 is called a visible o.p. triangle. Then Proposition 1 can be restated as follows: An opposed parastichy pair (m, n) is visible if and only if it has a visible o.p. triangle. Thus the problem of finding visible opposed parastichy pairs is reduced to the problem of finding visible o.p. triangles.

Proposition 2

If (m, n) is a visible opposed parastichy pair, then the number of fundamental spirals is equal to the greatest common divisor of m and n (Bravais, 1837). In what follows we assume that there is only one fundamental spiral, with leaves numbered in order using any leaf as leaf 0.

Characteristics of a visible o.p. pair

For a visible opposed parastichy pair (m, n), the visible o.p. triangle belonging to it has these characteristics. The base is 00_1. There are m steps on the left leg and n steps on the right leg. The first leaf on the left leg above 0 is leaf n. The first leaf on the right leg above 0_1 is leaf m. The apex is leaf mn.

(D) CONTRACTION OF A VISIBLE O.P. TRIANGLE

Definition

Let OPO_1 be a visible o.p. triangle belonging to the visible opposed parastichy pair (m, n), with m or n greater than 1. One of the two numbers m and n is greater than the other, since they are relatively prime. Suppose $m > n$. Let Q be the leaf on segment OP that is n steps from P. Triangle OQO_1 is called the *contraction* of triangle OPO_1. We shall call the opposed parastichy pair $(m-n, n)$ the *contraction* of the visible opposed parastichy pair (m, n).

Proposition 3

The contraction of a visible o.p. pair is visible.

Proposition 4

Every opposed parastichy pair $(a, 1)$ or $(1, a)$ that has an o.p. triangle with base 1 is visible.

(E) EXTENSION OF A VISIBLE O.P. TRIANGLE

The process of contraction of a visible o.p. triangle can be reversed. If OPO_1 is a visible o.p. triangle belonging to the opposed parastichy pair (m, n), extend O_1P through P m steps to Q, or extend OP through P n steps to Q'.

Definition

Triangle OQO_1 and triangle $OQ'O_1$ are called *extensions* of the visible o.p. triangle OPO_1. Note that while a visible o.p. triangle has only one contraction, it has two extensions. We call them left and right extensions, respectively, depending on whether the left leg or the right leg was extended. However, an extension of a visible o.p. triangle is not necessarily an o.p. triangle, since the legs may be both right parastichies or both left parastichies, instead of one right and one left.

Proposition 5

If an extension of a visible o.p. triangle is an o.p. triangle, then it is visible. If a left extension of a visible o.p. triangle belonging to the opposed parastichy pair (m, n) is an o.p. triangle, and hence visible, then it belongs to the opposed parastichy pair $(m+n, n)$. If a right extension is visible, it belongs to the opposed parastichy pair $(m, m+n)$.

In what follows we assume that the fundamental spiral is a right spiral.

Definition

If an opposed parastichy pair (m, n) is the final result of k successive extensions of a visible opposed parastichy pair (p, q), let us call (m, n) an *extension of order k* of (p, q). Let us also call (p, q) an extension of order 0 of itself.

Proposition 6

If (m, n), with $m, n > 1$, is a visible opposed parastichy pair, then (m, n) is an extension of some order of a unique visible opposed parastichy pair $(t, t+1)$, with $t > 1$.

Proposition 7

If the visible opposed parastichy pair (m, n) is an extension of some order of the pair $(t, t+1)$, then $1/(t+1) \leqslant d \leqslant 1/t$.

Proposition 8

Let (m, n) be a visible opposed parastichy pair with $m, n > 1$. If $1/(t+1) < d < 1/t$, (m, n) is an extension of some order of $(t, t+1)$. If $d = 1/t$, with $t > 2$, (m, n) is an extension of some order of either $(t, t+1)$ or $(t-1, t)$. If $d = 1/2$, (m, n) is an extension of some order of either $(2, 3)$ or $(3, 2)$.

Theorem 1 is the fundamental theorem on visibility of extensions:

THEOREM 1

If the closed segment of real numbers $[x/y, z/w]$ with $0 < x/y < z/w \leqslant 1/2$, and the parastichy triangle OPO_1 have the property that

(1) *the left leg of the triangle has w steps, each with a projection of length $-x+yd$ on OO_1; the right leg of the triangle has y steps each with a projection of length $z-wd$ on OO_1;*

(2) *the triangle is a visible o.p. triangle if and only if $x/y \leqslant d \leqslant z/w$,*

then the segment $[x/y, (x+z)/(y+w)]$ and the left extension of triangle OPO_1, and the segment $[(x+z)/(y+w), z/w]$ and the right extension of triangle OPO_1 have the same property.

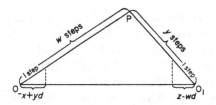

FIG. 7. Hypothesis for Theorem 1: The opposed parastichy triangle OPO_1, with the properties shown in the figure, is visible if and only if $x/y \leqslant d \leqslant z/w$.

Proposition 9

The closed segment $[1/(t+1), 1/t]$ and the o.p. triangle of the parastichy pair $(t, t+1)$ satisfy the hypothesis of Theorem 1.

Definition

If $x/y < z/w$, and both fractions are in lowest terms, then $(x+z)/(y+w)$ is called their *mediant* (Le Veque, 1956). It is easy to see that $x/y < (x+z)/(y+w) < z/w$.

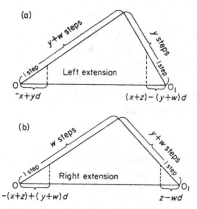

FIG. 8. Conclusion of Theorem 1: The left extension (a) of triangle OPO_1 is visible if and only if $x/y \leqslant d \leqslant (x+z)/(y+w)$; the right extension (b) is visible if and only if $(x+z)/(y+w) \leqslant d \leqslant z/w$.

Definition

A mediant subdivision of order k of the closed interval $[1/(t+1), 1/t]$ is a sequence S_k of subintervals defined inductively as follows:

(1) S_1 is the sequence of two closed subintervals into which $[1/(t+1), 1/t]$ is divided by the mediant of its endpoints, with the subintervals arranged in their natural order.

$$S_1 = [1/(t+1), 2/(2t+1)], [2/(2t+1), 1/t].$$

(2) S_{k+1} is the sequence of subintervals obtained from S_k by dividing each interval in S_k into closed subintervals by means of the mediant of its endpoints, and arranging them all in their natural order. Obviously S_k is a sequence of 2^k intervals.

We can assign a number to each member U_k of S_k as follows: Let $U_0 = [1/(t+1), 1/t]$. There is a unique sequence of intervals $U_1, U_2, \ldots, U_{k-1}$ such that $U_i \in S_i, i = 1, \ldots, k$, and $U_1 \supset U_2 \supset \ldots \supset U_{k-1} \supset U_k$. Assign to U_k the number whose binary numeral is defined as follows: It has k bits, and the ith bit from the left is 0 or 1 according as U_i is the left or right member of the mediant subdivision of U_{i-1}. It is easy to show that this numbers the members of S_k from left to right with the numbers $0, 1, \ldots, 2^k - 1$.

Definition

For a mediant subdivision of U_0 of a given order k, let us denote by $I_{k,j}$ the jth member of S_k.

Definition

An extension of order k of the opposed parastichy pair $(t, t+1)$ is obtained by making k extensions in succession, each of which is either a left extension or a right extension of the preceding one. For fixed k, the sequence of extensions can be assigned a number from 0 to $2^k - 1$ whose binary numeral is defined as follows: It has k bits, and the ith bit from the left is 0 or 1 according as the ith extension is left or right.

Definition

We denote by $E_{k,j}$ the extension of order k that is obtained by using the sequence of k extensions that has the number j.

THEOREM 2

$E_{k,j}$ *is visible if and only if* $d \in I_{k,j}$.

Another formulation of this result will also be useful. Let $n_k = b_1 b_2 \ldots b_k$ be a binary numeral with the k bits $b_1, b_2, \ldots, b_k$. We define a sequence of k binary numerals inductively as follows: $n_1 = b_1, n_i = n_{i-1}b_i, i = 2, \ldots, k$, where $n_{i-1}b_i$ represents a binary numeral whose first $i-1$ bits are those of n_{i-1} taken in order, and whose ith bit is b_i. Theorem 2 states that for any fixed k and n_k, E_{i,n_i} is visible if and only if

$$d \in I_{i, n_i}, i = 2, \ldots, k.$$

In view of the fact that

$$U_0 \supset I_{1, n_1} \supset I_{2, n_2} \supset \ldots \supset I_{k, n_k},$$

we can assert.

THEOREM 2'

For an fixed k and n_k, all the extensions E_{i, n_i} ($i = 1, 2, \ldots, k$) are visible if and only if

$$d \in \bigcap_{i=1}^{k} I_{i, n_i} = I_{k, n_k}.$$

INFINITE SEQUENCES OF EXTENSIONS

Let $b_1, b_2, \ldots, b_i, \ldots$ be an infinite sequence of bits. We can define an infinite sequence of binary numerals inductively as in the preceding paragraph: $n_1 = b_1, n_i = n_{i-1}b_i, i = 2, \ldots$. Then we have Theorem 3.

THEOREM 3

For any infinite sequence of bits b_1, b_2, ..., b_i, ..., all the extensions $E_{i,\,n_i}$ are visible if and only if

$$d \in \bigcap_{i=1}^{\infty} I_{i,\,n_i}.$$

THEOREM 3'

For any infinite sequence of bits b_1, b_2, ..., b_i, ..., all the extensions $E_{i,\,n_i}$ are visible if and only if d is the unique number in the nest of intervals $(I_{i,\,n_i})$.

RATIONAL DIVERGENCES

Proposition 10

The following propositions are equivalent:

(1) There exists a visible opposed parastichy pair one of whose members is a set of orthostichies.

(2) There exists a visible opposed parastichy pair both of whose extensions are visible.

(3) There are exactly two infinite sequences of successive visible extensions, each starting with an opposed parastichy pair of the form $(t, t+1)$, but not necessarily with the same value of t.

(4) If $(E_{i,\,n_i})$ is an infinite sequence of visible extensions of $(t, t+1)$ defined by the infinite sequence of bits (b_i), then there exists an index j such that either $b_i = 0$ for all $i > j$, or $b_i = 1$ for all $i > j$. That is, all but a finite number are extensions of the same kind, either all left or all right.

(5) There exist positive integers k and t, with $t \geqslant 2$, such that the divergence d is an endpoint of a member of S_k, the mediant subdivision of order k of the closed interval $[1/(t+1), 1/t]$.

(6) The divergence is rational.

IRRATIONAL DIVERGENCES

Proposition 10 can be restated as follows:

Proposition 10'

The following propositions are equivalent:

(1) There is no visible opposed parastichy pair one of whose members is a set of orthostichies.

(2) There is no visible opposed parastichy pair both of whose extensions are visible.

(3) There is exactly one infinite sequence of successive visible extensions starting with an opposed parastichy pair of the form $(t, t+1)$.

(4) Every infinite sequence of successive visible extensions of $(t, t+1)$ contains infinitely many left extensions and infinitely many right extensions.

(5) The divergence is irrational.

ALTERNATING SEQUENCES

Definition

Let (m, n) be a visible opposed parastichy pair that is an extension of order k of $(t, t+1)$. Then it is an extension of order $h = k+1$ of $(t, 1)$, the contraction of $(t, t+1)$. $(t, t+1)$ is a right extension of $(t, 1)$. If all the remaining extensions in the sequence of h successive visible extensions that ends with (m, n) are alternately left and right, the sequence is called an *alternating sequence* of order h. (The sequence is indexed with h in preference to k because it simplifies the notation in what follows.)

THEOREM 4

If (m, n) is a visible extension of order k of $(t, t+1)$, and it is the last term of an alternating sequence of order $h = k+1$ of visible extensions, then

(1) *if h is odd,*

$$m = S_{t, h}, n = S_{t, h+1},$$

and

$$F_{h+1}/S_{t, h+1} \leqslant d \leqslant F_h/S_{t, h};$$

(2) *if h is even,*

$$m = S_{t, h+1}, n = S_{t, h},$$

and

$$F_h/S_{t, h} \leqslant d \leqslant F_{h+1}/S_{t, h+1}.$$

THEOREM 4'

If a visible opposed parastichy pair consists of two consecutive terms of the sequence $(S_{t, h})$, then it is the last term of an alternating sequence that begins with either $(t, t+1)$ or $(t+1, t)$.

In the first case the fundamental spiral is a right-hand spiral and Theorem 4 applies. In the second case the fundamental spiral is a left-hand spiral and the corresponding version of Theorem 4 applies.

When $t = 2$, Theorem 4 becomes Theorem 5.

THEOREM 5

If (m, n) is a visible extension of order k of $(2, 3)$, and it is the last term of an alternating sequence of order $h = k+1$ of visible extensions, then
(1) *if h is odd,*
$$m = F_{h+2}, n = F_{h+3},$$
and
$$F_{h+1}/F_{h+3} \leqslant d \leqslant F_h/F_{h+2};$$
(2) *if h is even,*
$$m = F_{h+3}, n = F_{h+2},$$
and
$$F_h/F_{h+2} \leqslant d \leqslant F_{h+1}/F_{h+3}.$$

The case $t = 3$ is also of importance because it arises in polypeptide chains (Frey-Wyssling, 1954). An alpha-helix, for example, has about 3·6 amino-acid residues per turn.

An *infinite sequence* of successive visible extensions of $(t, t+1)$ is called *alternating* if the extensions are alternately left and right. In this case we have the following result.

THEOREM 6

If there exists an infinite alternating sequence of successive visible extensions of $(t, t+1)$, then the divergence is
$$\lim_{h \to \infty} F_h/S_{t,h} = (t+\tau^{-1})^{-1}.$$
If $t = 2$, the divergence is simply τ^{-2}.

APPENDIX C

Conspicuous Parastichies in Fibonacci Phyllotaxis

(1) Problem—to determine precisely the conditions under which a given visible opposed parastichy pair is conspicuous when $d = (t+\tau^{-1})^{-1}$.

By Theorem 4, successive visible extensions of the visible opposed parastichy pair $(t, t+1)$ all have the form $(S_{t,h}, S_{t,h+1})$ or $(S_{t,h+1}, S_{t,h})$. Then the first leaves up from leaf 0 on the legs of the corresponding visible o.p. triangle are $S_{t,h+1}$ and $S_{t,h}$, or vice versa (see Fig. 3 in section 5). By Corollary 4 of Theorem 7, these are consecutive points of close return. As the rise r decreases, a conspicuous set of parastichies determined by a point of close return

n_{k-2} gives way in conspicuousness to the parastichies determined by n_k. The transition from one set of parastichies to the other takes place when r has the value at which n_{k-2} and n_k are equidistant from leaf 0. Let us denote this value by r_k. Then

$$r_k^2 n_{k-2}^2 + (t+\tau^{-1})^{-2}\tau^{-2k+6} = r_k^2 n_k^2 + (t+\tau^{-1})^{-2}\tau^{-2k+2}.$$

$$r_k^2 S_{t,k-3}^2 + (t+\tau^{-1})^{-2}\tau^{-2k+6} = r_k^2 S_{t,k-1}^2 + (t+\tau^{-1})^{-2}\tau^{-2k+2}.$$

Then

$$r_k = \frac{(t+\tau^{-1})^{-1}\tau^{-k+1}(\tau^4-1)^{\frac{1}{2}}}{(S_{t,k-1}^2 - S_{t,k-3}^2)^{\frac{1}{2}}}.$$

When $t = 2$,

$$r_k = \frac{\tau^{-k-1}(\tau^4-1)^{\frac{1}{2}}}{(F_{k+1}^2 - F_{k-1}^2)^{\frac{1}{2}}} = \frac{(\tau+\tau^{-1})^{\frac{1}{2}}}{\tau^k(F_{2k})^{\frac{1}{2}}} = \frac{5^{\frac{1}{2}}}{\tau^k(F_{2k})^{\frac{1}{2}}}.$$

The parastichies determined by n_k are conspicuous when r is in the interval $[r_{k+2}, r_k]$. (Note that r_k decreases as k increases.) Then the opposed parastichy pair (n_{k-1}, n_k) or (n_k, n_{k-1}), whichever it happens to be, is conspicuous when

$$r \in [r_{k+1}, r_{k-1}] \cap [r_{k+2}, r_k],$$

or

$$r \in [r_{k+1}, r_k].$$

As a typical value of r when the opposed parastichy pair (n_{k-1}, n_k) or (n_k, n_{k-1}) is conspicuous we may take $\bar{r}_k = (r_k r_{k+1})^{\frac{1}{2}}$.

$$\bar{r}_k = \frac{5^{\frac{1}{2}}}{\tau^{k+\frac{1}{2}}(F_{2k}F_{2k+2})^{\frac{1}{2}}}.$$

$F_{2k} = F_k L_k$. Then

$$\bar{r}_k = \frac{5^{\frac{1}{2}}}{\tau^{k+\frac{1}{2}}(F_k F_{k+1} L_k L_{k+1})^{\frac{1}{2}}}.$$

Since

$$5F_k^2 = L_k^2 - 4(-1)^k, \quad L_k \doteq 5^{\frac{1}{2}} F_k.$$

Then

$$\bar{r}_k \doteq \tau^{-k-\frac{1}{2}}(F_k F_{k+1})^{-\frac{1}{2}}. \tag{C1}$$

This value for $\bar{r}_k$ turns out to be precisely the value of r for which the opposed parastichies determined by n_{k-1} and n_k are orthogonal (Coxeter, 1961).

When $r = r_k$, three sets of parastichies are conspicuous, namely, those determined by n_{k-2}, n_{k-1} and n_k. In this case, if the florets are in contact and are deformed by contact pressure, each is hexagonal, as is often the case in a pineapple. When $r = \bar{r}_k$, two sets of parastichies are conspicuous. In this case, if the florets are in contact and are deformed by contact pressure, each is rectangular.

The formula for $\bar{r}_k$ is an exact formula when the divergence is τ^{-2}. It is also useful as an approximate formula when the divergence is close to τ^{-2}.

(2) To verify that formula (C1) derived in a cylindrical representation agrees with the corresponding formula derived by Richards for a centric representation:

The Richards formula (1951, p. 517) for the orthogonal conspicuous opposed parastichy pair (a, b) is

$$\ln R_{(a,\, b)} = \left(\frac{\Psi_a \Psi_b}{ab}\right)^{\frac{1}{2}}. \tag{C2}$$

If $a = F_k$ and $b = F_{k+1}$, ψ_a and ψ_b are given by

$$\Psi_a = \tau^{-1}(2\pi)\tau^{-k+1} = 2\pi\tau^{-k},$$
$$\Psi_b = \tau^{-1}(2\pi)\tau^{-k} = 2\pi\tau^{-k-1}.$$

If we denote $R_{(a,\, b)}$ in this case by R_k, we have

$$\ln R_k = 2\pi\tau^{-k-\frac{1}{2}}(F_k F_{k+1})^{-\frac{1}{2}}. \tag{C3}$$

However, $R = e^{2\pi r}$. Hence $\ln R = 2\pi r$, and $\ln R_k = 2\pi r_k$. Therefore

$$2\pi r_k = 2\pi\tau^{-k-\frac{1}{2}}(F_k F_{k+1})^{-\frac{1}{2}},$$

and

$$r_k = \tau^{-k-\frac{1}{2}}(F_k F_{k+1})^{-\frac{1}{2}},$$

which is the same as formula (C1).

APPENDIX D

Producing Normal Phyllotaxis

In what follows the state of phyllotaxis of a leaf distribution is represented as a point in the (d, r) plane. As T increases and r decreases, the point (d, r) describes a path in the plane. We want to determine the nature of this path under conditions that we shall specify.

(1) Assume $1/3 < d \leqslant 1/2$, and only leaves 0 and 1 are present. For each value of d, max $\delta = $ dist $(0, 1)$. Let us write $f = \text{dist}^2 (0, 1) = d^2 + r^2$, and $f' = (1-d)^2 + r^2$. (f' is the distance from leaf 0 to leaf 1 measured to the left of 0, or the long way around the cylinder.) With r fixed, f and f' are functions of d. The graphs of these two functions are parabolas, both concave up, with vertices at $(0, r^2)$ and $(1, r^2)$ respectively (see Fig. 9).

Both f and f' are upper bounds for δ^2. Let us consider what may happen between the times $T = 1$ and $T = 2$.

(A) If throughout this interval of time $\delta^2 < f$, there is no contact pressure tending to push the centers of leaves 0 and 1 apart. Hence, as r falls, d may

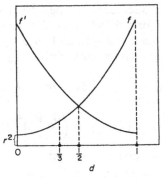

FIG. 9.

remain constant or may vary due to causes that are unrelated to contact pressure. Then the point (d, r) in the (d, r) plane moves down either on a vertical line or on some undetermined path.

(B) If at some time between $T = 1$ and $T = 2$ δ attains the maximum value determined by the value of d at that time, then at that time $\delta^2 = f$. If, moreover, δ continues to grow larger in spite of the decrease in r, we see from Fig. 9 that it can do so only if d increases, confirming our intuition that contact pressure, tending to push the centers of leaves 0 and 1 apart, would increase the horizontal component of the distance between them. Since f' as well as f is an upper bound for δ^2, δ^2 will attain its maximum possible value at $d = 1/2$. As d increases, the point (d, r) will move to the right in the (d, r) plane at the same time that it moves down with decreasing r. How fast it moves to the right depends on the rate of increase of d, which in turn depends on the rate of growth of δ. How fast it moves down depends on the rate of decrease of r. Then the path that the point (d, r) follows depends on the changing ratio of the rate of growth of δ and the rate of decrease of r.

Once d attains the value $1/2$, then, if δ remains maximized as r continues to fall, d will remain constant. Then the path followed by the point (d, r) in the (d, r) plane will no longer be indeterminate. It will be the vertical line $d = 1/2$.

(2) Assume again $1/3 < d \leqslant 1/2$. Let us consider now what may happen between the times $T = 2$ and $T = 3$, that is, after leaf 2 emerges but before leaf 3 emerges. As long as dist $(0, 2) >$ dist $(0, 1)$, leaves 0 and 1 continue to be in contact, and the argument outlined in (1) above continues to apply. (Note, however, that the presence of leaf 2 introduces one new relationship: as long as leaves 0 and 1 are in contact, so are leaves 1 and 2.) The picture changes, however, when r becomes so small that dist $(0, 2) \leqslant$ dist $(0, 1)$,

because then, if δ is maximized, leaves 0 and 2 are in contact, and if dist $(0, 2) <$ dist $(0, 1)$, the pressure on leaf 0 is exerted by leaf 2 and not by leaf 1. Let us redefine f and define g as follows: $f =$ dist^2 $(0, 2)$, $g =$ dist^2 $(0, 1)$. The horizontal component of dist $(0, 2)$ is $1 - 2d$, and the vertical component is $2r$. So we have

$$f(d) = (1 - 2d)^2 + 4r^2,$$
$$g(d) = d^2 + r^2. \tag{D1}$$

The condition for leaf 2 to exert contact pressure on leaf 0 is $f \leqslant g$. That is,

$$(1 - 2d)^2 + 4r^2 \leqslant d^2 + r^2, \tag{D2}$$

or

$$(d - 2/3)^2 + r^2 \leqslant (1/3)^2. \tag{D3}$$

This condition can be expressed graphically in the (d, r) plane in relation to the semicircle

$$(d - 2/3)^2 + r^2 = (1/3)^2, \qquad r > 0, \tag{D4}$$

as follows: Leaf 2 can exert contact pressure on leaf 0 if and only if the point (d, r) is on or under the semicircle defined by equation (D4). Since this semicircle is the locus of points (d, r) for which dist $(0, 2) =$ dist $(0, 1)$, we shall call it the $(2, 1)$ semicircle.

To see how contact pressure between leaves 0 and 2 may affect the value of d now, the graphs of the two smallest upper bounds for δ^2 are plotted. f and g are upper bounds for δ^2. Two other relevant upper bounds are f' and g' obtained from f and g respectively by replacing d by $1 - d$. Since $[1 - 2(1 - d)]^2 = (1 - 2d)^2$, $f' = f$, and since, when $d \leqslant 1/2, (1 - d)^2 \geqslant d^2$ so that $g' \geqslant g$, the two lowest upper bounds for δ^2 in the interval $0 < d \leqslant 1/2$ are f and g. Similarly, in the interval $1/2 \leqslant d < 1$, the two lowest upper bounds are f' and g'. The vertex of the graph of f is at $(1/2, 4r^2)$. $f \leqslant g$ for at least one point in the interval $1/3 < d \leqslant 1/2$ if and only if $f(1/2) \leqslant g(1/2)$, or

$$r \leqslant (1/6)\sqrt{3}. \tag{D5}$$

The two possibilities, $r < (1/6)\sqrt{3}$, and $r = (1/6)\sqrt{3}$, are shown in Figs 10 and 11.

For those values of d for which $g \leqslant f$, max $\delta^2 = g$. For those values of d for which $f \leqslant g$, max $\delta^2 = f$. Consequently, for each value of d in the interval $1/3 < d \leqslant 1/2$, max $\delta^2 = \min(f, g)$. Figs 10 and 11 show that this minimum is itself maximized in the interval at that value of d for which $f = g$ (The Maximin Criterion). Thus max δ^2, which is a function of d when r is fixed, has a maximum in the interval $1/3 < d \leqslant 1/2$ if r satisfies (D5).

Therefore, if r satisfies equation (D5) and δ is as large as possible, $\delta^2 = f = g$, and the point (d, r) must be on the $(2, 1)$ semicircle defined by equation (D4). The center of this semicircle is at $(2/3, 0)$ and its radius is

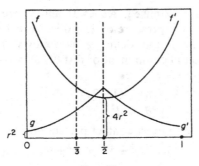

Fig. 10.

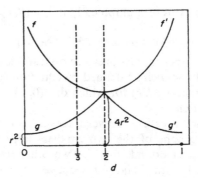

Fig. 11.

1/3. The interval $1/3 < d \leqslant 1/2$ lies entirely inside the interval $1/3 < d < 2/3$ in which equation (D4) defines d as a monotonic increasing function of r. So, as r continues to decrease, d must also decrease, with the point (d, r) moving to the left on the (2, 1) semicircle. From equation (D4) we find that on the (2, 1) semicircle, when $r < 1/5$, $d < 2/5$. So after r has fallen sufficiently, d will be restricted to the interval $1/3 < d < 2/5$.

While (d, r) is on the (2, 1) semicircle, $\delta^2 = f = g$, where f and g are defined by (D1), and r^2 is determined by equation (D3). Eliminating r^2 from equation (D3) and the equation for either f or g we find that on the (2, 1) semicircle

$$\delta^2 = (4/3)d - 1/3. \tag{D6}$$

Consequently, as d decreases with r, so does δ.

If $\delta^2 = \min (f, g)$ when r satisfies (D5), and if δ^2 is tending toward maximization but is not yet maximized, then the point (d, r) must move toward the (2, 1) semicircle as r falls, until it reaches the semicircle. We can see from

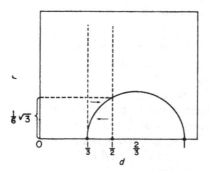

FIG. 12.

Figs 10, 11 and 12 that if the value of d is such that $f > g$, then d must increase for δ^2 to be maximized, and the point (d, r) must move to the right. If the value of d is such that $f < g$, then d must decrease, and the point (d, r) must move to the left. After the point (d, r) reaches the (2, 1) semicircle, it is constrained to move on it as described above.

We now examine what may happen as more and more leaves emerge, assuming that δ tends to remain maximized all the time. With δ maximized while the point (d, r) is on the (2, 1) semicircle, leaf 2 exerts contact pressure on leaf 0. As an abbreviation for the expression "leaf N exerts contact pressure on leaf 0" we shall say hereafter "leaf N pushes".

Pushing by leaves 1 and 2 dominates the situation until another leaf N emerges such that dist $(0, N)$ can be less than or equal to dist $(0, 2)$ for sufficiently small r. The first leaf that can satisfy this condition is the next point of close return after 2. In the interval $1/3 < d < 1/2$, leaf 3 is the next close return candidate after leaf 2. It will become the next point of close return when $d < 2/5$, and this condition will hold when $r < 1/5$, as we have seen.

Without specifying yet when leaf 3 emerges, let us determine the conditions under which leaf 3 can begin to push, and deduce the consequences of pushing by leaf 3. To do so, we make the same kind of analysis we made to find out the conditions for and the consequences of pushing by leaf 2.

It is assumed that leaves 0, 1, 2 and 3 are present and that $1/3 < d < 2/5$. Redefining f and g, $f = $ dist^2 $(0, 3)$, $g = $ dist^2 $(0, 2)$. In the notation of section 7, $n_2 = 2$ and $n_3 = 3$; $D_2 = 1 - 2d$, and $D_3 = 3d - 1$. Consequently,

$$f(d) = (3d - 1)^2 + 9r^2,$$
$$g(d) = (1 - 2d)^2 + 4r^2. \tag{D7}$$

The condition for leaf 3 to push is $f \leqslant g$, which becomes

$$(d - 1/5)^2 + r^2 \leqslant (1/5)^2. \tag{D8}$$

Therefore leaf 3 can push only if the point (d, r) is on or under the semicircle defined by

$$(d - 1/5)^2 + r^2 = (1/5)^2, \qquad r > 0. \tag{D9}$$

Since, when (d, r) is on this semicircle, dist $(0, 3) =$ dist $(0, 2)$, we shall call it the $(3, 2)$ semicircle.

When (d, r) is on or near the $(3, 2)$ semicircle, let us determine the conditions that are necessary for δ to be maximized. To do so, we first plot the graphs of f and g as functions of d, with r fixed. The graph of f is a parabola with vertex at $(1/3, 9r^2)$. The graph of g is a parabola with vertex at $(1/2, 4r^2)$. $g(1/2) < f(1/2)$. If we could be assured that $f((1/3) < g(1/3)$, then the parabolas would cross in the interval $1/3 < d < 1/2$, and min (f, g) would be maximized at the intersection point, where $f = g$. The condition that $f(1/3) < g(1/3)$ is

$$r < (1/15)\sqrt{5}. \tag{D10}$$

The point (d, r), moving on the $(2, 1)$ semicircle, enters the region where $f \leqslant g$ only after it reaches the $(3, 2)$ semicircle. Let us denote by (d_4, r_4) the co-ordinates of the intersection of these two semicircles, since this is the point at which leaf $F_4 = 3$ would first qualify to push. Then

$$d_4 = 5/14, \qquad r_4 = (1/14)\sqrt{3}. \tag{D11}$$

We may summarize what we have found out so far as follows: (1) leaf 3 becomes the next point of close return when $r < 1/5$; (2) leaf 3 can begin to push when $r \leqslant (1/14)\sqrt{3}$; (3) when leaf 3 pushes, maximization of δ compels (d, r) to be on the $(3, 2)$ semicircle if $r < (1/15)\sqrt{5}$. But $(1/14)\sqrt{3} < (1/15)\sqrt{5} < 1/5$, so once leaf 3 is present and pushing, maximization of δ requires that the point (d, r) be on the $(3, 2)$ semicircle.

Now we shall take into account the time when leaf 3 emerges, assuming that it emerges while (d, r) is on the $(2, 1)$ semicircle. There are three separate cases to be considered.

(1) Leaf 3 has already emerged, and its diameter becomes equal to dist $(0, 2)$ precisely when the point (d, r) is at (d_4, r_4), the intersection of the $(2, 1)$ semicircle and the $(3, 2)$ semicircle. Then the point (d, r) switches from the $(2, 1)$ semicircle to the $(3, 2)$ semicircle and moves on the latter as r continues to decrease.

(2) Leaf 3 emerges and its diameter becomes equal to dist $(0, 2)$ before (d, r) reaches the $(3, 2)$ semicircle. It cannot push until (d, r) reaches the $(3, 2)$ semicircle. So (d, r) will continue to move on the $(2, 1)$ semicircle until it reaches (d_4, r_4), and then it will switch to the $(3, 2)$ semicircle, *provided that before it switches no leaf $N > 3$ emerges that is capable of pushing before the switch takes place.* We postpone until later consideration of the question of

whether or how this condition can be satisfied. Meanwhile let us assume that it is satisfied.

(3) Leaf 3 may emerge before or after (d, r) crosses the (3, 2) semicircle, but its diameter does not fill the available space (equal to dist /0, 2) before the crossing, and equal to dist (0, 3) after the crossing) until after the crossing. Then δ is not maximized when this happens, and there is room for growth. As δ tends toward maximization the point (d, r) must move from the (2, 1) semicircle to the (3, 2) semicircle on some transitional path. The point (d, r) will be on the transitional path as long as δ at the value of r attained is not the maximum possible with that value of r. The nature of the transitional path and the length of time that the point (d, r) will be on it will depend on the rate of increase of δ and the rate of decrease of r. The point (d, r) will ultimately reach the (3, 2) semicircle, *provided that the maximization of δ proceeds fast enough, and provided that no leaf $N > 3$ diverts it by beginning to push before the (3, 2) semicircle is reached.* After the point (d, r) reaches the (3, 2) semicircle, it moves on it. It can be seen that the third and second cases introduce an element of uncertainty about the outcome. How this uncertainty can be eliminated is dealt with later in this section.

Whenever the point (d, r) makes the transition to the (3, 2) semicircle, $d > 1/3$, so d is in the interval $1/5 < d < 2/5$ in which equation (D9) defines d as a monotonic decreasing function of r. Thus, after the transition, d increases as r decreases, i.e. the point (d, r) moves to the right on the (3, 2) semicircle.

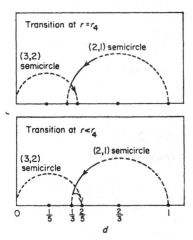

FIG. 13. Two possible types of transition from the (2, 1) semicircle to the (3, 2) semicircle.

Thus there are two types of transition of the point (d, r) from the $(2, 1)$ semicircle to the $(3, 2)$ semicircle (see Fig. 13). Note that either transition entails a reversal of the direction of change of d.

While (d, r) is on the $(3, 2)$ semicircle, $\delta^2 = f = g$ defined by equation (D7), while r^2 is given by equation (D9). Eliminating r^2 from equations (D7) and (D9), we find that on the $(3, 2)$ semicircle

$$\delta^2 = 1 - (12/5)d. \tag{D12}$$

On the $(3, 2)$ semicircle we will have $d > 3/8$ when $r < (1/40)\sqrt{15}$.

While (d, r) is on the $(2, 1)$ semicircle, the phyllotaxis of the plant is $(2, 1)$ phyllotaxis. During the transition to the $(3, 2)$ semicircle, and while (d, r) is on the $(3, 2)$ semicircle, the phyllotaxis is $(2, 3)$ phyllotaxis.

The type of analysis we have already used in this section can be generalized by induction as follows:

Theorem 8

For $n \geqslant 4$, assume that δ is maximized before $T = 3$, that r is decreasing, that d is between F_{n-2}/F_n and F_{n-1}/F_{n+1}, and that the point (d, r) is on the (F_{n-1}, F_{n-2}) semicircle, with leaves F_{n-1} and F_{n-2} pushing. The next leaf that will qualify to push is leaf F_n. f and g are redefined as follows:

$$f(d) = \text{dist}^2(0, F_n) = (F_n d - F_{n-2})^2 + F_n^2 r^2,$$
$$g(d) = \text{dist}^2(0, F_{n-1}) = (F_{n-1} d - F_{n-3})^2 + F_{n-1}^2 r^2. \tag{D13}$$

The vertex of the graph of f is at $(F_{n-2}/F_n, F_n^2 r^2)$, and the vertex of the graph of g is at $(F_{n-3}/F_{n-1}, F_{n-1}^2 r^2)$. The condition $f = g$ defines the (F_n, F_{n-1}) semicircle whose equation is

$$\left(d - \frac{F_{n-2}F_{n-1} + (-1)^{n-1}}{F_{n-2}F_{n+1}}\right)^2 + r^2 = (F_{n-2}F_{n+1})^{-2}, \qquad r > 0. \tag{D14}$$

For fixed r, $\min(f, g)$ is maximized when $f = g$ at a value of d between F_{n-2}/F_n and F_{n-3}/F_{n-1} if

$$f(F_{n-2}/F_n) < g(F_{n-2}/F_n),$$

or

$$r < [F_n(F_{n-2}F_{n+1})^{\frac{1}{2}}]^{-1}. \tag{D15}$$

Denote by (d_n, r_n) the intersection of the (F_n, F_{n-1}) semicircle and the (F_{n-1}, F_{n-2}) semicircle. Then

$$d_n = \frac{4F_{n-2}^2 + (-1)^n}{2(2F_{n-1}^2 + (-1)^{n-1})}, \tag{D16}$$

$$r_n = \frac{\sqrt{3}}{2(2F_{n-1}^2 + (-1)^{n-1})}.$$

Leaf F_n is the next point of close return after leaf F_{n-1} when $r < s_n$, where

$$s_n = \frac{\sqrt{3}}{F_{n+1}\sqrt{F_{n-3}F_n}}, \qquad \text{(D17)}$$

and it is able to push when $r < r_n$. Moreover,

$$r_n < (F_n(F_{n-2}F_{n+1})^{\frac{1}{2}})^{-1},$$

so that when F_n is able to push, maximization of δ requires that $f = g$, or that (d, r) be on the (F_n, F_{n-1}) semicircle. Then if leaf F_n is present and fully grown [its diameter equal to dist $(0, F_{n-1})$] *when $r \geqslant r_n$, the point (d, r) switches from the (F_{n-1}, F_{n-2}) semicircle to the (F_n, F_{n-1}) semicircle at their inter-section* provided that before it switches no leaf $N > F_n$ emerges that is capable of pushing before the switch takes place. *If leaf F_n is present and fully grown [its diameter equal to* dist $(0, F_n)$] *only after $r < r_n$, then, as δ tends toward maximization, (d, r) moves along a transitional path to the (F_n, F_{n-1}) semi-circle, and will ultimately reach it* provided that the maximization of δ proceeds fast enough, and provided that no leaf $N > F_n$ diverts it by begin-ning to push before the (F_n, F_{n-1}) semicircle is reached. *After either type of transition to the (F_n, F_{n-1}) semicircle, the point (d, r) moves to the right on it if n is even, and moves to the left on it if n is odd, so that the transition initiates a reversal of direction. When*

$$r < \frac{\sqrt{3}}{F_{n+2}\sqrt{F_{n-2}F_{n+1}}}, \qquad \text{(D18)}$$

d will lie between F_{n-1}/F_{n+1} and F_n/F_{n+2}. While (d, r) is on the (F_n, F_{n-1}) semicircle, the phyllotaxis is (F_n, F_{n-1}) or (F_{n-1}, F_n) according as n is odd or even, and if δ_n is the (maximized) leaf diameter during this time,

$$\delta_n^2 = \frac{(-1)^n}{F_{n-2}F_{n+1}} \left[2F_{n-2}F_{n-1} + (-1)^n - 2dF_nF_{n-1}\right]. \qquad \text{(D19)}$$

The proof is computational, and is based on Theorem 7 and some of the identities in Appendix A.

For $n \geqslant 3$, equations (D16) define a sequence of points $P_n = (d_n, r_n)$. For $n \geqslant 3$, each pair of consecutive points (P_n, P_{n+1}) is joined by an arc of the (F_n, F_{n-1}) semicircle, and the sequence of these arcs forms a continuous curve that zig-zags back and forth as it proceeds from one P_n to the next. This curve is called *the normal phyllotaxis path* (for $t = 2$).

Theorem 8 states that if the relative leaf diameter is maximized before $T = 3$, and if (d, r) is on the arc of the normal phyllotaxis path joining

P_{n-1} and P_n with leaves F_{n-1} and F_{n-2} pushing, it may make the transition to the next arc, joining P_n and P_{n+1}, in two ways: (1) by switching to the next arc as soon as it reaches it, if leaf F_n is present and fully grown by then, or (2) by leaving the normal phyllotaxis path temporarily and then rejoining it, if leaf F_n emerges and is fully grown only after (d, r) has passed through the point P_n. The second type of transition can be eliminated entirely by specifying that every leaf F_n emerges and becomes fully grown while (d, r) is between P_{n-1} and P_n, so that $r > r_n$. Then the transition will take place when P_n is reached, provided that no leaf $N > F_n$ begins pushing before that time. We can assure that this condition is satisfied by specifying that every leaf F_n emerges only after $r < s_n$:

THEOREM 9

If (1) *r is decreasing,* (2) *the relative leaf diameter δ is maximized before* $T = 3$, *and* (3) *for every integer* $n > 3$ *leaf* F_n *emerges when*

$$r_n \leqslant r < s_n, \tag{D20}$$

then from the time leaf 2 is fully grown the point (d, r) *is always on the normal Fibonacci path, and d approaches* τ^{-2} *as a limit as* $n \to \infty$.

To prove this theorem we need only show that, when (d, r) is on the arc of the (F_{n-1}, F_{n-2}) semicircle joining P_{n-1} and P_n, no leaf emerging after leaf F_n can push before F_n does. By hypothesis, leaf F_{n+1} emerges after $r < s_{n+1}$, and it is easily verified that $s_{n+1} < r_n$, so that leaf F_{n+1} and later leaves emerge only after leaf F_n is already pushing. Leaves that emerge after F_n but before F_{n+1} cannot push before F_n does, because while $r_n < r < s_n$ d is in a range of values where F_{n+1} is the next close return candidate after F_n, so that for $F_n < N < F_{n+1}$ dist $(0, N) >$ dist $(0, F_n)$. That d approaches τ^{-2} as a limit follows from the fact that d_n is between F_{n-2}/F_n and F_{n-1}/F_{n+1}.

What happens under the conditions of Theorem 9 is shown in Fig. 7 in section 11, and in Tables 1 and 2. Note that in Table 1, if the headings of columns 3, 4, 5, 6 and 7 are read consecutively, they form a sentence.

Condition (D20) of Theorem 9 is satisfied for all $n > 3$ if the rate of emergence of new leaves and the rate of decrease of r are such that

$$r = 1 \cdot 5 T^{-2}. \tag{D21}$$

Proof

Table 3 shows the values of r obtained from equation (D21) for values of $T = F_n$ for $n = 4, \ldots, 12$.

TABLE 1

How the phyllotaxis of a plant changes as r decreases if $1/3 < d < 1/2$, leaf diameter is always as great as possible, and leaf F_n emerges when

$$\sqrt{3}(2(2F_{n-1}^2 + (-1)^{n-1}))^{-1} \leqslant r < \sqrt{3}(F_{n+1}\sqrt{F_{n-3}F_n})^{-1}$$

(1)	(2)	(3)		(4)		(5)	(6)	(7)
		If F_n emerges when		then when r is in		the phyllotaxis		and the range
n	F_n	r is in this interval,		this inteval,		is (F_n, F_{n-1}),	the range of d is,	of δ^2 is
3	2			0·1237	$\leqslant r <$ 0·2887	(2, 1)	0·35714 $\leqslant d <$ 0·50000	0·1428571 $\leqslant \delta^2 <$ 0·3333333
4	3	0·1237	$\leqslant r <$ 0·2000	0·0456	$\leqslant r \leqslant$ 0·1237	(3, 2)	0·35714 $\leqslant d \leqslant$ 0·39474	0·0526315 $\leqslant \delta^2 \leqslant$ 0·1428571
5	5	0·0456	$\leqslant r <$ 0·0968	0·0177	$\leqslant r \leqslant$ 0·0456	(5, 3)	0·37755 $\leqslant d \leqslant$ 0·39474	0·0204081 $\leqslant \delta^2 \leqslant$ 0·0526315
6	8	0·0177	$\leqslant r <$ 0·0333	0·0067	$\leqslant r \leqslant$ 0·0177	(8, 5)	0·37755 $\leqslant d \leqslant$ 0·38372	0·0077519 $\leqslant \delta^2 \leqslant$ 0·0204081
7	13	0·0067	$\leqslant r <$ 0·0132	0·00257	$\leqslant r \leqslant$ 0·0067	(13, 8)	0·38131 $\leqslant d \leqslant$ 0·38372	0·0029673 $\leqslant \delta^2 \leqslant$ 0·0077519
8	21	0·00257	$\leqslant r <$ 0·00497	0·00098	$\leqslant r \leqslant$ 0·00257	(21, 13)	0·38131 $\leqslant d \leqslant$ 0·38222	0·0011325 $\leqslant \delta^2 \leqslant$ 0·0029673
9	34	0·00098	$\leqslant r <$ 0·00191	0·00037	$\leqslant r \leqslant$ 0·00098	(34, 21)	0·38187 $\leqslant d \leqslant$ 0·38222	0·0004327 $\leqslant \delta^2 \leqslant$ 0·0011325
10	55	0·00037	$\leqslant r <$ 0·00073	0·00014	$\leqslant r \leqslant$ 0·00037	(55, 34)	0·38187 $\leqslant d \leqslant$ 0·38200	0·0001652 $\leqslant \delta^2 \leqslant$ 0·0004327
11	89	0·00014	$\leqslant r <$ 0·00028	0·000055	$\leqslant r \leqslant$ 0·00014	(89, 55)	0·38195 $\leqslant d \leqslant$ 0·38200	0·0000631 $\leqslant \delta^2 \leqslant$ 0·0001652
12	144	0·000055	$\leqslant r <$ 0·000106	0·000021	$\leqslant r \leqslant$ 0·000055	(144, 89)	0·38195 $\leqslant d \leqslant$ 0·38197	0·0000241 $\leqslant \delta^2 \leqslant$ 0·0000631

I. ADLER

TABLE 2

How δ^2 varies with d as the phyllotaxis of a plant evolves if $1/3 < d < 1/2$ and Leaf F_n emerges when

$$\sqrt{3}(2(2F_{n-1}^2+(-1)^{n-1})) \leqslant r < \sqrt{3}(F_{n+1}\sqrt{F_{n-3}F_n})^{-1}$$

Phyllotaxis	δ^2 as a function of d
(2, 1)	$(1/3)(4d-1)$
(3, 2)	$(1/5)(5-12d)$
(5, 3)	$(1/16)(30d-11)$
(8, 5)	$(1/39)(31-80d)$
(13, 8)	$(1/105)(208d-79)$
(21, 13)	$(1/272)(209-546d)$
(34, 21)	$(1/715)(1428d-545)$
(55, 34)	$(1/1869)(1429-3740d)$
(89, 55)	$(1/4896)(9790d-3739)$
(144, 89)	$(1/12815)(9791-25632d)$

TABLE 3

n	$T = F_n$	r
4	3	0·1667
5	5	0·0600
6	8	0·02344
7	13	0·00888
8	21	0·00340
9	34	0·0012975
10	55	0·0004958
11	89	0·0001893
12	144	0·0000723

A comparison of Table 3 with Table 1 shows that a plant whose r is governed by equation (D21) for $3 \leqslant T \leqslant 144$ does indeed satisfy condition (D20). In fact, condition (D20) is still satisfied for $3 \leqslant T \leqslant 144$ if (D21) is generalized to

$$1\cdot14 \leqslant r(T)\cdot T^2 < 1\cdot79 \tag{D22}$$

where $r(T)$ is the value of r at time T. Moreover, with equation (D22), condition (D20) is satisfied even for $T > 144$. To prove this fact, let $F_n > 144$. By identity (A11) of Appendix A, for large n, $F_n \doteq \tau^n 5^{-\frac{1}{2}}$. So at time $T = F_n$, $F_{n-1} \doteq T\tau^{-1}$, $F_{n+1} \doteq T\tau$, and $F_{n-3} \doteq T\tau^{-3}$. Substituting these values into condition (D20), keeping in mind the definition of r_n and s_n given by equation (D16) and equation (D17), we find that condition (D20) becomes

$$1 \cdot 1335 T^{-2} \leqslant r \leqslant 2 \cdot 203 T^{-2}, \tag{D23}$$

which is clearly satisfied by equation (D22). Thus we have the following result:

THEOREM 10 (MAIN THEOREM)

If the relative leaf diameter δ of the leaf primordia is maximized at time $T_c < 3$, and if the rise at time T is governed by equation (D22) for $3 \leqslant T \leqslant F_N$, then (d, r) is on the normal phyllotaxis path throughout the interval of time $T_c \leqslant T \leqslant F_N$, and the plant develops higher and higher normal phyllotaxis at least through (F_{N-1}, F_{N-2}) phyllotaxis.

It must be stressed that the condition that F_n emerge and be fully grown when r satisfies condition (D20) in Theorem 9 is a sufficient condition to produce normal phyllotaxis, but it is not a necessary condition. In fact, it is possible to prove for $n = 4$, 5 and 6 that, when (d, r) is on the (F_{n-1}, F_{n-2}) semicircle, as long as F_n emerges when $r > r_n$, no leaf that emerges after leaf F_n can push before leaf F_n does, because the higher the leaf number, the smaller the value of r at which it can begin to push. Thus there are conditions more general than those given by Theorem 9 that can produce normal phyllotaxis.

In Theorem 8 we have taken $t = 2$. That is, we have assumed that $1/3 < d < 1/2$. The theorem is easily extended to the more general situation $1/(t+1) < d < 1/t, t \geqslant 2$, as follows:

THEOREM 11

For $t \geqslant 2$, $n \geqslant 2$, assume that δ is maximized before $T = 3$, that r is decreasing, that d is between $F_n/S_{t,n}$ and $F_{n+1}/S_{t,n+1}$, and that the point (d, r) is on the $(S_{t,n-1}, S_{t,n-2})$ semicircle with leaves $S_{t,n-1}$ and $S_{t,n-2}$ pushing. The next leaf that will qualify to push is leaf $S_{t,n}$.

Define

$$f(d) = \text{dist}^2 (0, S_{t,n}) = (S_{t,n}d - F_n)^2 + S_{t,n}^2 r^2 ,$$
$$g(d) = \text{dist}^2 (0, S_{t,n-1}) = (S_{t,n-1}d - F_{n-1})^2 + S_{t,n-1}^2 r^2 . \tag{D24}$$

The vertex of the graph of f is at $(F_n/S_{t,n}, S_{t,n}^2 r^2)$, and the vertex of the graph of g is at $(F_{n-1}/S_{t,n-1}, S_{t,n-1}^2 r^2)$. The condition $f = g$ defines the $(S_{t,n}, S_{t,n-1})$

semicircle whose equation is

$$\left(d - \frac{F_{n-2}S_{t,n+1}+(-1)^n}{S_{t,n-2}S_{t,n+1}}\right)^2 + r^2 = (S_{t,n-2}S_{t,n+1})^{-2}, \qquad r > 0. \tag{D25}$$

For fixed r, min (f, g) is maximized when f = g at a value of d between $F_n/S_{t,n}$ and $F_{n-1}/S_{t,n-1}$ if $f(F_n/S_{t,n}) < g(F_n/S_{t,n})$, or

$$r < [S_{t,n}(S_{t,n-2}S_{t,n+1})^{\frac{1}{2}}]^{-1}. \tag{D26}$$

Denote by (d_n, r_n) the intersection of the $(S_{t,n}, S_{t,n-1})$ semicircle and the $(S_{t,n-1}, S_{t,n-2})$ semicircle. Then

$$d_n = \frac{4F_{n-1}S_{t,n-1}+(2t-1)(-1)^{n-1}}{2(2S_{t,n-1}^2+(-1)^{n-1}(t^2-t-1))}, \tag{D27}$$

$$r_n = \frac{\sqrt{3}}{2(2S_{t,n-1}^2+(-1)^{n-1}(t^2-t-1))}.$$

Leaf $S_{t,n}$ is the next point of close return after leaf $S_{t,n-1}$ when $r < s_n$, where

$$s_n = \frac{\sqrt{3}}{S_{t,n+1}\sqrt{S_{t,n-3}S_{t,n}}}, \tag{D28}$$

and it is able to push when $r < r_n$.

Moreover, $r_n < (S_{t,n}(S_{t,n-2}S_{t,n+1})^{\frac{1}{2}})^{-1}$, so that, when $S_{t,n}$ is able to push, maximization of δ requires that f = g, or that (d, r) be on the $(S_{t,n}, S_{t,n-1})$ semicircle. Then if leaf $S_{t,n}$ is present and fully grown [its diameter equal to dist $(0, S_{t,n-1})$] when $r \geqslant r_n$, the point (d, r) switches from the $(S_{t,n-1}, S_{t,n-2})$ semicircle to the $(S_{t,n}, S_{t,n-1})$ semicircle at their intersection, provided that before it switches no leaf $N > S_{t,n}$ emerges that is capable of pushing before the switch takes place. If leaf $S_{t,n}$ is present and fully grown [its diameter equal to dist $(0, S_{t,n})$] only after $r < r_n$, then, as δ tends toward maximization, (d, r) moves along a transitional path to the $(S_{t,n}, S_{t,n-1})$ semicircle, and will ultimately reach it, provided that the maximization of δ proceeds fast enough and provided that no leaf $N > S_{t,n}$ diverts it by beginning to push before the $(S_{t,n}, S_{t,n-1})$ semicircle is reached. After either type of transition to the $(S_{t,n}, S_{t,n-1})$ semicircle, the point (d, r) moves to the right on it if n is even, and moves to the left on it if n is odd, so that the transition entails a reversal of direction. When,

$$r < \frac{\sqrt{3}}{S_{t,n+2}\sqrt{S_{t,n-2}S_{t,n+1}}}, \tag{D29}$$

d will be between $F_{n+1}/S_{t,n+1}$ and $F_{n+2}/S_{t,n+2}$. While (d, r) is on the $(S_{t,n},$

$S_{t,n-1}$) semicircle, the phyllotaxis is $(S_{t,n}, S_{t,n-1})$ or $(S_{t,n-1}, S_{t,n})$ according as n is odd or even, and if δ_n is the (maximized) leaf diameter during this time,

$$\delta_n^2 = \frac{(-1)^n}{S_{t,n-2}S_{t,n+1}}(2F_{n-1}S_{t,n}+(-1)^{n+1}-2dS_{t,n}S_{t,n-1}).$$

Theorem 11 reduces to Theorem 8 when $t = 2$, with a slight change of notation, namely replacing n by $n-2$, to take into account the fact that $S_{2,n} = F_{n+2}$.

For $n \geqslant 1$ equation (D27) defines a sequence of points $P_n = (d_n, r_n)$. Each pair of consecutive points (P_n, P_{n+1}) is joined by an arc of the $(S_{t,n}, S_{t,n-1})$ semicircle, and the sequence of these arcs forms a continuous curve that zig-zags back and forth as it proceeds from one P_n to the next. Let us call this curve *the normal phyllotaxis path* (for given t). Then Theorem 9 can be generalized for $t \geqslant 2$ as follows:

THEOREM 12

If (1) *r is decreasing,* (2) *the relative leaf diameter is maximized before* $T = t+1$, *and* (3) *for every integer* $n > 1$ *leaf* $S_{t,n}$ *emerges when* $r_n \leqslant r < s_n$, *where* r_n *is defined by equation* (D27) *and* s_n *is defined by equation* (D28), *then, from the time leaf t is fully grown, the point* (d, r) *is always on the normal Fibonacci path, and d approaches* $(t+\tau^{-1})^{-1}$ *as a limit as* $n \to \infty$.

APPENDIX E

Producing Anomalous Phyllotaxis

We assume to begin with that $1/3 < d < 1/2$, and that the maximization of the relative leaf diameter δ is first attained at a time T such that $5 < T < 6$. Then before leaf 5 emerges, the point (d, r) follows a path in the (d, r) plane that is undetermined by contact pressure. After leaf 5 emerges, and δ is maximized, the continuation of this path is determined by contact pressure. To identify conditions under which this continuation of the path of (d, r) produces anomalous phyllotaxis, we first consider which of the leaves 1, 2, 3, 4 and 5 may be able to exert contact pressure on leaf 0. The first step is to determine dist (l_0, n) for $n = 1, \ldots, 5$. To do so we calculate nd, $d(n)$ and $c(n)$. Then dist $(l_0, n) = d(n)$ or $c(n)$, whichever is smaller. The results of this calculation are shown in Table 4.

Now we can use the formulas for dist (l_0, n) to determine which pairs of leaves selected from leaves 1, 2, 3, 4 and 5 may be nearest to leaf 0 and thus be able to exert contact pressure on it. These five leaves determine 10 pairs

TABLE 4

nd	$d(n)$	$c(n)$	dist (l_0, n)
$1/3 < d < 1/2$	$1/3 < d(1) < 1/2$	$1/2 < c(1) < 2/3$	$d(1) = d$
$2/3 < 2d < 1$	$2/3 < d(2) < 1$	$0 < c(2) < 1/3$	$c(2) = 1{-}2d$
$1 < 3d < 3/2$	$0 < d(3) < 1/2$	$1/2 < c(3) < 1$	$d(3) = 3d{-}1$
$4/3 < 4d < 2$	$1/3 < d(4) < 1$	$0 < c(4) < 2/3$	$d(4) = 4d{-}1$ if $d \leqslant 3/8$
			$c(4) = 2{-}4d$ if $d \geqslant 3/8$
$5/3 < 5d < 5/2$	$2/3 < d(5) \leqslant 1$	$0 \leqslant c(5) < 1/3$	$c(5) = 2{-}5d$ if $d \leqslant 2/5$
	$0 \leqslant d(5) < 1/2$	$1/2 < c(5) \leqslant 1$	$d(5) = 5d{-}2$ if $d \geqslant 2/5$

altogether, but we must exclude the four pairs containing leaf 4 for the following reason. If $d \geqslant 3/8$, dist $(l_0, 4) = 2-4d = 2$ dist $(l_0, 2)$. Then dist $(0, 4) = 2$ dist $(0, 2)$. If either $(4, 1)$, $(4, 3)$ or $(4, 5)$ are the pairs of leaves nearest to leaf 0, then dist $(0, 2) \geqslant$ dist $(0, 4)$, which is impossible. If $(4, 2)$ is the pair of leaves nearest to leaf 0, leaf 4 can exert contact pressure only if there exist values of d for which dist $(0, 4) \leqslant$ dist $(0, 2)$, which is not possible. If $d < 3/8$, dist $(l_0, 4) = 4d-1$. Then leaf 4 can exert contact pressure on leaf 0 only if $4d-1 \leqslant d$, or $4d-1 \leqslant 1-2d$, or $4d-1 \leqslant 3d-1$, or $4d-1 \leqslant 2-5d$. The third of these conditions requires $d \leqslant 0$, and the others require $d \leqslant 1/3$, contradicting our assumption that $d > 1/3$. Therefore we need consider only the leaf pairs $(2, 1)$, $(3, 1)$, $(3, 2)$, $(5, 1)$, $(5, 2)$ and $(5, 3)$. Each of these pairs (m, n) with $m > n$ determines the (m, n) semi-circle defined by dist^2 $(0, m) =$ dist^2 $(0, n)$, $r > 0$, with the property that dist $(0, m)$ is less than, equal to, or greater than dist $(0, n)$ according as (d, r) is under the semicircle, or on it, or outside it. The six semicircles are shown in Fig. 5, section 12. They divide the plane into regions, and for each region we can determine which leaf pair is nearest to leaf 0 when (d, r) is in that region. For example, region I is outside the $(5, 2)$ semicircle and the $(3, 2)$ semicircle, but is under the others. Therefore, for (d, r) in region I,

$$\text{dist } (0, 5) > \text{dist } (0, 2), \text{dist } (0, 3) > \text{dist } (0, 2),$$
$$\text{dist } (0, 5) < \text{dist } (0, 3), \text{dist } (0, 5) < \text{dist } (0, 1),$$
$$\text{dist } (0, 3) < \text{dist } (0, 1), \text{ and dist } (0, 2) < \text{dist } (0, 1).$$

Then

$$\text{dist } (0, 1) > \text{dist } (0, 3) > \text{dist } (0, 5) > \text{dist } (0, 2),$$

so that leaves 5 and 2 are nearest to leaf 0. Similarly, region II is outside only the $(3, 2)$ semicircle and is under the others. Therefore, for (d, r) in region II,

$$\text{dist } (0, 3) > \text{dist } (0, 2), \text{dist } (0, 5) < \text{dist } (0, 2),$$
$$\text{dist } (0, 5) < \text{dist } (0, 3), \text{dist } (0, 5) < \text{dist } (0, 1),$$
$$\text{dist } (0, 3) < \text{dist } (0, 1), \text{ and dist } (0, 2) < \text{dist } (0, 1).$$

Then

$$\text{dist } (0, 1) > \text{dist } (0, 3) > \text{dist } (0, 2) > \text{dist } (0, 5),$$

so that leaves 5 and 2 are nearest to leaf 0. It is easily verified that these are the only regions in which leaves 5 and 2 are nearest to leaf 0.

Now we shall determine what may happen if δ first becomes maximized while (d, r) is in region I or II. The equation of the $(5, 2)$ semicircle is

$$r^2 + (d - 8/21)^2 = (1/21)^2, \qquad r > 0. \tag{E1}$$

On this semicircle, values of d are confined to the interval $1/3 < d < 3/7$, whose midpoint is $d = 8/21$. Define f and g by

$$f(d) = \text{dist }^2(0, 5) = (5d - 2)^2 + 25r^2,$$
$$g(d) = \text{dist}^2(0, 2) = (2d - 1)^2 + 4r^2. \tag{E2}$$

The vertices of the graphs of f and g are at $d = 2/5$ and $d = 1/2$ respectively. $\min (f, g)$ has a maximum in the interval $2/5 < d < 1/2$ if $f(2/5) < g(2/5)$. This condition requires that $r < \sqrt{21}/105 \doteq 0\cdot0436435$. On the right-hand half of the $(5, 2)$ semicircle, $r < \sqrt{21}/105$ if and only if $d > 2/5$. Moreover, we see from Table 4 that leaves 2 and 5 are on opposite sides of l_0 so that $(2, 5)$ is an opposed parastichy pair if and only if $d \geqslant 2/5$. Let R be the rectangle whose base is the interval $2/5 < d < 1/2$ on the d-axis, and whose height is $\sqrt{21}/105$. Then if δ begins to become maximized when (d, r) is in $R \cap (I \cup II)$, the point (d, r) moves to the $(5, 2)$ semicircle and then moves on it as r decreases. It moves to the right on the $(5, 2)$ semicircle because $2/5 > 8/21$, and in the interval $8/21 < d < 3/7$ (E1) defines d as a monotonic decreasing function of r. In the range $2/5 < d < 3/7$, nd, $d(n)$ and $c(n)$ for $n = 1, \ldots, 7$ have the values shown in Table 5.

It can be seen from Table 5 that the first three points of close return are $n_1 = 1, n_2 = 2$ and $n_3 = 5$. Moreover, leaf 7 is the next close return candidate

TABLE 5

nd	$d(n)$	$c(n)$
$2/5 < d < 3/7$	$2/5 < d(1) < 3/7$	$4/7 < c(1) < 3/5$
$4/5 < 2d < 6/7$	$4/5 < d(2) < 6/7$	$1/7 < c(2) < 1/5$
$6/5 < 3d < 9/7$	$1/5 < d(3) < 3/7$	$4/7 < c(3) < 4/5$
$8/5 < 4d < 12/7$	$3/5 < d(4) < 5/7$	$2/7 < c(4) < 2/5$
$2 < 5d < 15/7$	$0 < d(5) < 1/7$	$6/7 < c(5) < 1$
$12/5 < 6d < 18/7$	$2/5 < d(6) < 4/7$	$3/7 < c(6) < 3/5$
$14/5 < 7d < 3$	$4/5 < d(7) < 1$	$0 < c(7) < 1/5$

after leaf 5 (for the interval $2/5 < d < 3/7$). Leaf 7 is the next point of close return n_4 if and only if $c(7) < d(5)$, or $3-7d < 5d-2$, or $d > 5/12$. On the $(5, 2)$ semicircle, $d > 5/12$ when $r < \sqrt{7}/84 \doteq 0\cdot031496$. So, as (d, r) moves on the $(5, 2)$ semicircle, leaf 7 becomes the next point of close return after leaf 5 when $r < \sqrt{7}/84$. f and g are redefined as follows:

$$f(d) = \text{dist}^2(0, 7) = (3-7d)^2+49r^2,$$
$$g(d) = \text{dist}^2(0, 5) = (5d-2)^2+25r^2. \tag{E3}$$

The vertices of the graphs of f and g are at $d = 3/7$ and $d = 2/5$ respectively. $\min(f, g)$ has a maximum in the interval $2/5 < d < 3/7$ if $f(3/7) < g(3/7)$, or $r < \sqrt{6}/84 \doteq 0\cdot0291605$. The $(7, 5)$ semicircle defined by $f = g, r > 0$, has the equation

$$r^2+(d-11/24)^2 = (1/24)^2. \tag{E4}$$

The $(5, 2)$ semicircle and the $(7, 5)$ semicircle intersect at $d = 11/26$, $r = \sqrt{3}/78 \doteq 0\cdot022057$. Therefore if leaf 7 emerges and is fully grown when $\sqrt{3}/78 \leqslant r < \sqrt{7}/84$, it is already the next point of close return after leaf 5. If the next close return candidate after leaf 7 emerges only after $r < \sqrt{3}/78$, no leaf after leaf 7 can push before leaf 7 does. Then, when (d, r) reaches the intersection of the $(5, 2)$ semicircle and the $(7, 5)$ semicircle, leaf 7 begins to push, and because $r < \sqrt{6}/84$, (d, r) must then move on the $(7, 5)$ semicircle. While (d, r) was on the $(5, 2)$ semicircle, the phyllotaxis was $(2, 5)$. After the transition to the $(7, 5)$ semicircle, the phyllotaxis becomes $(7, 5)$. Thus we have identified a set of conditions that will produce anomalous phyllotaxis first in the form of $(2, 5)$ phyllotaxis and later in the form of $(7, 5)$ phyllotaxis. The conditions are that (1) δ was first maximized when $5 \leqslant T < 6$; (2) when δ is first maximized (d, r) is in the region $R \cap (I \cup II)$; (3) leaf 7 emerges and is fully grown when $\sqrt{3}/78 \leqslant r < \sqrt{7}/84$, and the next close return candidate after leaf 7 emerges when $r < \sqrt{3}/78$.

APPENDIX F

Convergence of Inhibitor-determined d_n

PROOF OF THEOREM A

(1) Let $H(x, y) = g(x)+ag(x+y)$. Assume $0 < y \leqslant 1/2$, and keep y fixed. $H(0, y) = -\infty$; $H(1/2, y) = -ag([1/2]-y) > 0$. $H(x, y)$ is a monotonic increasing function of x in the interval $0 \leqslant x \leqslant 1/2$ and changes sign in that interval. Therefore, for every y in the interval $0 \leqslant y \leqslant 1/2$, there is a unique

value of x in the interval $0 < x < 1/2$ such that $H(x, y) = 0$. Thus the recurrence relation (31) determines a unique sequence (d_n) with $0 < d_n < 1/2$ for $n > 1$.

(2) $H(x, y)$ is a monotonic increasing function of y in the interval $0 < y \leqslant 1/2$. Consider the function $\bar{H}(x) = H(x, x)$. $\bar{H}(0) = -\infty$; $\bar{H}(1/2) = \infty$. $\bar{H}(x)$ is a monotonic increasing function of x in the interval $0 < x < 1/2$ and changes sign in that interval. Therefore there is a unique root $x = d$ such that $\bar{H}(d) = H(d, d) = 0$, and $0 < d < 1/2$.

(3) We now show that the sequence (d_n) converges to d. Suppose $d_{n-1} < d$. Then, since $H(d, d) = 0$, and $H(x, y)$ is a monotonic increasing function of y, we have $H(d, d_{n-1}) < 0$. However, $H(d_n, d_{n-1}) = 0$. Therefore $d_n > d$. Similarly, if $d_{n-1} > d$, $d_n < d$. Since $d_1 = 1/2 > d$, $d_{2i-1} > d$, $d_{2i} < d$ for $i = 1, 2, 3, \ldots$. We now show that the sequence (d_{2i-1}) is monotonic decreasing, and that the sequence (d_{2i}) is monotonic increasing.

$$H(d_{n-2}, d_{n-1}) = g(d_{n-2}) + ag(d_{n-2} + d_{n-1}).$$

$$0 = H(d_{n-1}, d_{n-2}) = g(d_{n-1}) + ag(d_{n-1} + d_{n-2}).$$

Therefore

$$H(d_{n-2}, d_{n-1}) = g(d_{n-2}) - g(d_{n-1}).$$

If n is odd,

$$d_{n-1} < d < d_{n-2}, \text{ and } g(d_{n-1}) < g(d_{n-2}),$$

hence

$$H(d_{n-2}, d_{n-1}) > 0.$$

But

$$H(d_n, d_{n-1}) = 0.$$

Therefore

$$d_n < d_{n-2}.$$

Similarly, if n is even,

$$d_n > d_{n-2}.$$

Thus

$$d_1 > d_3 > d_5 > \ldots > d,$$

$$d_2 < d_4 < d_6 < \ldots < d.$$

Then the monotonic sequence (d_{2i-1}), $i = 1, 2, 3, \ldots$, approaches some limit $\alpha \geqslant d$, and the monotonic sequence (d_{2i}), $i = 1, 2, 3, \ldots$, approaches some limit $\beta \leqslant d$. Then

$$0 = H(\alpha, \beta) = g(\alpha) + ag(\alpha + \beta),$$

and

$$0 = H(\beta, \alpha) = g(\beta) + ag(\beta + \alpha).$$

Then

$$g(\alpha) = g(\beta).$$

Since $g(x)$ is monotonic in the interval

$$0 < x < 1, \alpha = \beta = d.$$

(4) Since $g(x)$ is continuous in the interval

$$0 < x < 1, 0 = H(d, d) = g(d) + ag(2d).$$

Therefore

$$a = -[g(d)]/[g(2d)].$$

If $d \leqslant 1/4$, $2d \leqslant 1/2$, $g(d) < 0$ and $g(2d) \leqslant 0$, and $a < 0$ or is infinite, which contradicts the assumption $a > 0$. Therefore $d > 1/4$.

PROOF OF THEOREM B

Let

$$H(x, y, z) = g(x) + ag(x+y) + bg(x+y+z),$$

where $0 < b \leqslant a$; (d_n) is defined as follows:

$$d_1 = 0 \cdot 5; \quad d_2 = \text{the root of } g(d_2) + ag(d_2 + 0 \cdot 5) = 0$$

with

$$0 < d_2 < 0 \cdot 5; \quad d_n \text{ is the root of } H(d_n, d_{n-1}, d_{n-2}) = 0$$

with

$$0 < d_n < 0 \cdot 5, \quad n = 3, 4, 5, \ldots.$$

(1) It is assumed that the sequence (d_n) exists and is well-defined, and

$$d_n + d_{n-1} + d_{n-2} < 1 \text{ for } n > 2.$$

(2) Suppose $d_n \leqslant d_{n-1} < d_{n-2}$. We prove that $d_{n+1} > d_n$:

$$0 = g(d_{n+1}) + ag(d_{n+1} + d_n) + bg(d_{n+1} + d_n + d_{n-1}).$$
$$0 = g(d_n) + ag(d_n + d_{n-1}) + bg(d_n + d_{n-1} + d_{n-2}).$$

If $d_{n+1} \leqslant d_n$, then $g(d_{n+1}) \geqslant g(d_n)$,

$$g(d_{n+1} + d_n) \geqslant g(d_n + d_{n-1}),$$

and

$$g(d_{n+1} + d_n + d_{n-1}) > g(d_n + d_{n-1} + d_{n-2}).$$

Hence

$$0 = H(d_{n+1}, d_n, d_{n-1}) > H(d_n, d_{n-1}, d_{n-2}) = 0,$$

which is a contradiction. Therefore $d_{n+1} > d_n$. A similar proof shows that if $d_n \geqslant d_{n-1} > d_{n-2}$, then $d_{n+1} < d_n$. Hence a sequence of at most three consecutive terms of (d_n) can be monotonic, and if d_n is the last term of such a maximal monotonic non-increasing sequence, $d_{n+1} > d_n$; if d_n is the last term of such a maximal non-decreasing sequence, $d_{n+1} < d_n$. Thus the sequence (d_n) is a succession of two-term or three-term maximal monotonic sequences in which the last term of each monotonic sequence is the first term of the next one, and the second term of each sequence is greater or less than the first term.

It is now shown that if one of these maximal monotonic sequences is non-decreasing, its last term is less than or equal to the first term of the maximal non-increasing sequence that precedes it, and if the sequence is non-increasing its last term is greater than or equal to the first term of the maximal non-decreasing sequence that precedes it. The four cases are considered that may arise as each of two successive maximal monotonic sequences may have either two terms or three terms.

(3) $d_2 < d_1 = 0\cdot5$. If $d_3 > d_2$, nevertheless, since $d_3 < 0\cdot5$, $d_3 < d_1$. For $n > 2$, suppose $d_n < d_{n-1}$, while $d_{n-1} \geqslant d_{n-2}$. We prove that if $d_{n+1} > d_n$, then $d_{n+1} \leqslant d_{n-1}$.

$$0 = g(d_{n+1}) + ag(d_{n+1}+d_n) + bg(d_{n+1}+d_n+d_{n-1}).$$
$$0 = g(d_n) + ag(d_n+d_{n+1}) + bg(d_n+d_{n-1}+d_{n-2}).$$

If $d_{n+1} > d_{n-1}$, then $g(d_{n+1}) < g(d_{n-1})$, and $g(d_{n+1}+d_n) < g(d_n+d_{n-1})$. Then $g(d_{n+1}+d_n+d_{n-1}) > g(d_n+d_{n-1}+d_{n-2})$, and hence $d_{n+1} < d_{n-2}$. But $d_{n+1} > d_{n-1} \geqslant d_{n-2}$ implies that $d_{n+1} > d_{n-2}$, so we have a contradiction. Hence $d_{n+1} \leqslant d_{n-1}$. A similar proof shows that for $n > 2$, if $d_n > d_{n-1}$, $d_{n-1} \leqslant d_{n-2}$, and $d_{n+1} < d_n$, then $d_{n+1} \geqslant d_{n-1}$.

(4) Suppose $d_n \leqslant d_{n-1} < d_{n-2}$. Then by (2), $d_{n+1} > d_n$. We prove that $d_{n+1} < d_{n-2}$. If $d_{n+1} \leqslant d_{n-1}$, then since $d_{n-1} < d_{n-2}$, we have $d_{n+1} < d_{n-2}$. If $d_{n+1} > d_{n-1}$, since $d_{n+1} > d_n$, $g(d_{n+1}) < g(d_n)$ and $g(d_{n+1}+d_n) < g(d_n+d_{n-1})$.

$$0 = g(d_n) + ag(d_n+d_{n-1}) + bg(d_n+d_{n-1}+d_{n-2}).$$
$$0 = g(d_{n+1}) + ag(d_{n+1}+d_n) + bg(d_{n+1}+d_n+d_{n-1}).$$

Therefore $g(d_{n+1}+d_n+d_{n-1}) > g(d_n+d_{n-1}+d_{n-2})$, and $d_{n+1} < d_{n-2}$. A similar proof shows that if $d_n \geqslant d_{n-1} > d_{n-2}$, then, although $d_{n+1} < d_n$, $d_{n+1} > d_{n-2}$.

(5) $d_2 < d_1 = 0\cdot5$. If $d_4 \geqslant d_3 > d_2$, nevertheless $d_4 < d_1$, since $d_4 < 0\cdot5$. For $n > 2$, suppose $d_n < d_{n-1}$ while $d_{n-1} \geqslant d_{n-2}$. We prove that if $d_{n+2} \geqslant d_{n+1} > d_n$, then $d_{n+2} \leqslant d_{n-1}$. By (3), $d_{n+1} \leqslant d_{n-1}$. So if $d_{n+2} = d_{n+1}$, $d_{n+2} \leqslant d_{n-1}$. Suppose $d_{n+2} > d_{n-1}$. Then $g(d_{n+2}) < g(d_{n+1})$. Also, since $d_{n+2} > d_n$, $g(d_{n+2}+d_{n+1}) < g(d_{n+1}+d_n)$.

$$0 = g(d_{n+2}) + ag(d_{n+2}+d_{n+1}) + bg(d_{n+2}+d_{n+1}+d_n).$$
$$0 = g(d_{n+1}) + ag(d_{n+1}+d_n) + bg(d_{n+1}+d_n+d_{n-1}).$$

Then $g(d_{n+2}+d_{n+1}+d_n) > g(d_{n+1}+d_n+d_{n-1})$. Hence $d_{n+2} < d_{n-1}$. A similar proof shows that if $d_n > d_{n-1}$, $d_{n-1} \leqslant d_{n-2}$ and $d_{n+2} \leqslant d_{n+1} < d_n$, then $d_{n+2} \geqslant d_{n-1}$.

(6) Suppose $d_n \leqslant d_{n-1} < d_{n-2}$, and $d_{n+2} \geqslant d_{n+1} > d_n$. We prove that $d_{n+2} < d_{n-2}$. By (4), $d_{n+1} < d_{n-2}$. Then if $d_{n+2} = d_{n+1}$, $d_{n+2} < d_{n-2}$. Suppose $d_{n+2} > d_{n+1}$. Then $g(d_{n+2}) < g(d_{n+1})$. Also, since $d_{n+2} > d_n$,

$g(d_{n+2}+d_{n+1}) < g(d_{n+1}+d_n)$. Then $g(d_{n+2}+d_{n+1}+d_n) > g(d_{n+1}+d_n+d_{n-1})$, and $d_{n+2} < d_{n-1}$. But $d_{n-1} < d_{n-2}$, so $d_{n+2} < d_{n-2}$. A similar proof shows that if $d_n \geqslant d_{n-1} > d_{n-2}$, and $d_{n+2} \leqslant d_{n+1} < d_n$, then $d_{n+2} > d_{n-2}$.

(7) The sequence (d_n) with $n > 1$ may therefore be partitioned into three subsequences (α_i), (β_i) and (γ_i), such that the α_i are the last terms of the maximal monotonic non-increasing sequences in (d_n); the β_i are the last terms of the maximal monotonic non-decreasing sequences in (d_n); and the γ_i are all the rest of the terms of (d_n). In each of the sequences (α_i), (β_i), and (γ_i) the terms are indexed to show the order of their appearance in (d_n). $\beta_i = d_1$. The terminal points of the nth non-increasing sequence are β_n, α_n with $\alpha_n < \beta_n$. The terminal points of the nth non-decreasing sequence are α_n, β_{n+1} with $\alpha_n < \beta_{n+1}$. We have shown above that $\alpha_{n+1} \geqslant \alpha_n$, and $\beta_{n+1} \leqslant \beta_n$ for $n = 1, 2, 3, \ldots$. $\alpha_n < \beta_1$, and $\beta_n > \alpha_i$ for $n = 1, 2, 3, \ldots$. Therefore the sequence (α_n) has a limit α, and the sequence (β_n) has a limit β, and $\alpha \leqslant \beta$.

Each maximal monotonic sequence $\beta_n, \ldots, \alpha_n$ and $\alpha_n, \ldots, \beta_{n+1}$ contains at most one term of (γ_i). $\alpha_1 \leqslant \gamma_i < \beta_1$. There are two cases to consider.

Case A

There are only finitely many γ_i. Then for sufficiently large n, three consecutive terms of (d_i) are of the form $(\beta_n, \alpha_n, \beta_{n+1})$ or $(\alpha_n, \beta_{n+1}, \alpha_{n+1})$. Since $g(x)$ is continuous for $0 < x < 1$,

$$0 = \lim_{n \to \infty} H(\beta_n, \alpha_n, \beta_{n+1}) = H(\beta, \alpha, \beta),$$

and

$$0 = \lim_{n \to \infty} H(\alpha_n, \beta_{n+1}, \alpha_{n+1}) = H(\alpha, \beta, \alpha).$$

Then

$$0 = g(\beta) + ag(\beta + \alpha) + bg(2\beta + \alpha),$$

and

$$0 = g(\alpha) + ag(\alpha + \beta) + bg(2\alpha + \beta).$$

Therefore

$$g(\beta) + bg(2\beta + \alpha) = g(\alpha) + bg(2\alpha + \beta).$$

If

$$\alpha < \beta, g(\beta) < g(\alpha)$$

and

$$g(2\beta + \alpha) < g(2\alpha + \beta),$$

and hence

$$g(\beta) + bg(2\beta + \alpha) < g(\alpha) + bg(2\alpha + \beta),$$

and we have a contradiction. Hence $\alpha = \beta$.

Case B

There are infinitely many γ_i. Then γ_i has at least one limit point. Let γ be such a limit point. There exists a subsequence (γ_{n_i}) whose limit is γ, and $\alpha \leqslant \gamma \leqslant \beta$. Each γ_{n_i} is a term of a maximal monotonic sequence in (d_n) whose other terms are members of (α_i) and (β_i) respectively. Considering now only those triples (d_n, d_{n-1}, d_{n-2}) one of whose terms is a term of (γ_i), there are three cases to consider:

    (a) the γ_{n_i} occur in only finitely many maximal non-increasing sequences $\beta_n, \ldots, \alpha_n$;

    (b) the γ_{n_i} occur in only finitely many maximal non-decreasing sequences $\alpha_n, \ldots, \beta_{n+1}$;

    (c) the γ_{n_i} occur in infinitely many maximal non-increasing sequences $\beta_n, \ldots, \alpha_n$ and in infinitely many maximal non-decreasing sequences $\alpha_n, \ldots, \beta_{n+1}$.

In case (a), for sufficiently large index the γ_{n_i} occur only in maximal non-decreasing sequences. Renumber the γ_{n_i} as γ_i. Then, for sufficiently large n, a triple (d_n, d_{n-1}, d_{n-2}), one of whose terms is one of these γ_i may be of one of these three types: $(\gamma_i, \alpha_{n_i}, \beta_{n_i})$, $(\beta_{n_i+1}, \gamma_i, \alpha_{n_i})$ and $(\alpha_{n_i+1}, \beta_{n_i+1}, \gamma_i)$ where α_{n_i} and β_{n_i+1} are the terminal points of the maximal non-decreasing sequence that contains γ_i. Then, letting i approach ∞,

$$0 = H(\gamma, \alpha, \beta) = H(\beta, \gamma, \alpha) = H(\alpha, \beta, \gamma).$$

That is,

$$0 = g(\gamma) + ag(\gamma + \alpha) + bg(\gamma + \alpha + \beta).$$
$$0 = g(\beta) + ag(\beta + \gamma) + bg(\beta + \gamma + \alpha).$$
$$0 = g(\alpha) + ag(\alpha + \beta) + bg(\alpha + \beta + \gamma).$$

From the last two equations,

$$g(\beta) + ag(\beta + \gamma) = g(\alpha) + ag(\alpha + \beta).$$

If $\gamma > \alpha$, $\beta > \alpha$, and $\beta + \gamma > \alpha + \beta$, and hence $g(\beta) + ag(\beta + \gamma) < g(\alpha) + ag(\alpha + \beta)$, which is a contradiction. Therefore $\gamma = \alpha$. Then we have

$$g(\alpha) + ag(2\alpha) = g(\alpha) + ag(\alpha + \beta).$$

Consequently $g(2\alpha) = g(\alpha + \beta)$, $2\alpha = \alpha + \beta$, and $\alpha = \beta$. In case (b) for sufficiently large index, the γ_{n_i} occur only in maximal non-increasing sequences. Renumbering these γ_{n_i} as γ_i, then, for sufficiently large n, triples (d_n, d_{n-1}, d_{n-2}) one of whose terms is one of these γ_i may be of one of these three types: $(\gamma_i, \beta_{n_i}, \alpha_{n_i-1})$, $(\alpha_{n_i}, \gamma_i, \beta_{n_i})$, and $(\beta_{n_i+1}, \alpha_{n_i}, \gamma_i)$, where α_{n_i} and β_{n_i} are the terminal points of the maximal non-increasing sequence that contains γ_i. Then

$$0 = H(\gamma, \beta, \alpha) = H(\alpha, \gamma, \beta) = H(\beta, \alpha, \gamma).$$

That is,

$$0 = g(\gamma) + ag(\gamma + \beta) + bg(\gamma + \beta + \alpha).$$
$$0 = g(\alpha) + ag(\alpha + \gamma) + bg(\alpha + \gamma + \beta).$$
$$0 = g(\beta) + ag(\beta + \alpha) + bg(\beta + \alpha + \gamma).$$

From the last two equations,

$$g(\alpha) + ag(\alpha + \gamma) = g(\beta) + ag(\beta + \alpha).$$

If $\gamma > \alpha$, $\beta > \alpha$, and $\beta + \alpha \geqslant \alpha + \gamma$, and hence

$$g(\alpha) + ag(\alpha + \gamma) > g(\beta) + ag(\beta + \gamma),$$

which is a contradiction. Therefore $\gamma = \alpha$. Then we have

$$g(\alpha) + ag(\alpha + \beta) = g(\alpha) + ag(2\alpha).$$

Consequently $\alpha = \beta$. In case (c), let (γ_h) be the subsequence of (γ_{n_i}) consisting of the γ_{n_i} that occur in a maximal non-increasing sequence $(\beta_{n_h}, \gamma_h, \alpha_{n_h})$; let (γ_k') be the subsequence of (γ_{n_i}) consisting of those γ_{n_i} that occur in a maximal non-decreasing sequence $(\alpha_{n_k}, \gamma_k', \beta_{n_k+1})$. Then letting h and k increase to infinity, we get

$$0 = H(\beta, \gamma, \alpha) = H(\alpha, \gamma, \beta).$$
$$0 = g(\beta) + ag(\beta + \gamma) + bg(\beta + \gamma + \alpha).$$
$$0 = g(\alpha) + ag(\alpha + \gamma) + bg(\alpha + \gamma + \beta).$$

Then

$$g(\beta) + ag(\beta + \gamma) = g(\alpha) + ag(\alpha + \gamma).$$

If $\beta > \alpha$, we have

$$g(\beta) + ag(\beta + \gamma) < g(\alpha) + ag(\alpha + \gamma),$$

which is a contradiction. Therefore $\beta = \alpha$. Thus in all cases we have $\alpha = \beta = \gamma$, and d_n converges to α.

(8) Now that we know that in the three-plastochrone case d_n converges to a limit d, we show that

$$1/5 < d < 1/3.$$

Since

$$0 < d_n + d_{n-1} + d_{n-2} < 1,$$

and

$$\lim_{n \to \infty} d_n = d, \ 0 \leqslant 3d \leqslant 1.$$
$$\bar{H}(d) = g(d) + ag(2d) + bg(3d) = 0.$$

If $d = 1/3$, $2d = 2/3$, $3d = 1$. Then g(1/3) and g(2/3) are finite, but g(1) is not. So 1/3 cannot be a root of $\bar{H}(d) = 0$. If $d = 0.2$, $2d = 0.4$, $3d = 0.6$.

$$\bar{H}(0.2) = g(0.2) + ag(0.4) + bg(0.6) = g(0.2) + ag(0.4) - bg(0.4) > 0,$$

since

$$g(0.2) > 0 \text{ and } 0 < b \leqslant a.$$

TABLE 6

Two plastochrone case

a	Value of d when			
	$r = 0$	$r = 0.01$	$r = 0.03$	$r = 0.07$
0·01	0·4576	0·4588	0·4697	—
0·03	0·4394	0·4402	0·4469	—
0·1	0·4113	0·4118	0·4160	0·442
0·3	0·3773	0·3776	0·3802	0·3948
0·6	0·3524	0·3526	0·3546	0·3649
0·9	0·3373	0·3375	0·3391	0·3474
1·0	0·3333	0·3336	0·3351	0·3429
2·0	0·3086	0·3087	0·3098	0·3151
3·0	0·2958	0·2959	0·2967	0·3007
4·0	0·288	0·288	0·288	0·292
5·0	0·282	0·282	0·283	0·285
10·0	0·269	0·27	0·27	0·270

TABLE 7

Three plastochrone case with $a \leqslant 1$

a	b	Value of d when			
		$r = 0$	$r = 0.01$	$r = 0.03$	$r = 0.07$
0·1	0·033	0·32	nv	nv	nv
0·1	0·067	0·31	(0·31)	nv	nv
0·1	0·1	0·30	(0·30)	nv	nv
0·3	0·1	0·298	0·30	nv	nv
0·3	0·2	0·286	0·287	nv	nv
0·3	0·3	0·278	0·279	nv	nv
0·6	0·2	0·285	0·286	nv	nv
0·6	0·4	0·271	0·272	(0·279)	nv
0·6	0·6	0·262	0·263	0·2688	nv
0·9	0·3	0·276	0·277	(0·2845)	nv
0·9	0·6	0·262	0·2625	0·2680	nv
0·9	0·9	0·253	0·2531	0·2576	nv
1·0	0·33	0·274	0·275	(0·2815)	nv
1·0	0·67	0·259	0·260	0·2652	nv
1·0	1·0	0·25	0·2505	0·2548	nv

nv equation not valid because $d_n + d_{n-1} + d_{n-2} > 1$ for some n.
( ) limit probably valid although $d_n + d_{n-1} + d_{n-2} > 1$ for some n.

If

$$0 < d < 0.2, 2d < 0.4, 3d < 0.6.$$

Then

$$g(2d) > g(0.4), \text{ and } g(3d) > g(0.6).$$

$$\bar{H}(d) = g(d) + ag(2d) + bg(3d) > g(d) + ag(0.4) + bg(0.6) >$$
$$g(d) + ag(0.4) - bg(0.4) > g(d) > 0.$$

If $d = 0$, $\bar{H}(d)$ is infinite. Therefore $\bar{H}(d) = 0$ requires $1/5 < d < 1/3$.

Tables 6, 7 and 8 summarize the results of computer calculations using equations (30) and (31) and $f(x, y) = 1/s + 1/s'$.

TABLE 8

Three plastochrone case with $a > 1$

a	b	Value of d when			
		$r = 0$	$r = 0.01$	$r = 0.03$	$r = 0.07$
2	0.5	0.264	0.264	0.269	(0.296)
2	1	0.250	0.250	0.254	
2	2	0.234	0.234	0.237	
3	1	0.250	0.250	0.254	(0.272)
3	2	0.235	0.236	0.238	(0.253)
3	3	0.226	0.226	0.229	(0.241)
4	1	0.250	0.250	0.253	(0.268)
4	2	0.236	0.237	0.239	
4	4	0.221	0.222	0.223	0.234
5	1	0.250	0.250	0.253	(0.266)
5	3	0.229	0.229	0.231	(0.243)
5	5	0.218	0.218	0.220	0.2296
10	3.3	0.2316	0.234	0.236	
10	6.7	0.2182	0.220	0.222	0.23
10	10	0.2100	0.210	0.212	0.219

(　) limit probably valid although $d_n + d_{n-1} + d_{n-2} > 1$ for some n.
(Blank spaces in the table indicate places where the time allotted to the computer program ran out before the computation was completed.)

APPENDIX G

Brief Outline of the History of Phyllotaxis

THEOPHRASTUS AND PLINY

Theophrastus (370 B.C.–285 B.C.) knew that there was an orderly arrangement of leaves in many plants. He said, for example, in his "Enquiry Into Plants" (Theophrastus, 1916), "And in general those that have flat leaves have them in a regular series, ..." Leaf arrangement was one of the characteristics used to help identify a plant. Thus Pliny (23 A.D.–79 A.D.) in his "Natural History," (1856, tr. by Bostock and Riley) described aparine as a "ramose, hairy plant with five or six leaves at regular intervals, arranged circularly around the branches."

THE QUINCUNX

The arrangement of leaves came to be known as a *quincunx* since it resembled the quincunx pattern of staggered rows of grape vines. The dictionary defines the pattern as "an arrangement of plants with one at each corner and one at the center of a rectangle or square." According to Church (1904), the *quinque* in the name refers to the V formation that is repeated rather than to the five trees in each square.

KEPLER

Johannes Kepler (1571–1630) was the first person to assert that the Fibonacci numbers (2) had a special significance in phyllotaxis. He pointed to the frequent occurrence of the number five in plants, as in "the arrangement of the leaves in a group of five" on trees, and in "the five limited spaces for holding the seeds" in an apple. He identified the number five as one of the terms of sequence (2) which "propagates itself" by the recurrence relation (1), and then said "I think that the seeding capacity of a tree is fashioned in a manner similar to the above sequence propagating itself." (Ludwig, 1896).

SAUVAGES AND LINNAEUS

Sauvages distinguished four different kinds of leaf arrangement: opposite leaves, whorls of three or more leaves, alternate leaves on opposite sides of the stem, and leaves with no constant arrangement. (Bonnet, 1754) Linnaeus (1707–1778) added several more types to this classification, but neither he nor Sauvages mentioned the existence of spirals.

DA VINCI AND BONNET

Botanical literature usually credits Charles Bonnet (1720–1793) with being the first to notice and study the spiral arrangement of leaves. He observed the common case in which the divergence is 2/5, and called it the quincunx, thus changing the meaning of this term. He introduced a separate classification of multiple spirals called to his attention by his friend Calandrini. From his descriptions and drawings it is clear that the multiple spirals are parastichies. He also formulated a theory to explain why leaves have a spiral arrangement: with a spiral arrangement the leaves cover each other as little as possible, thus allowing the air to circulate freely among the leaves (Bonnet, 1754).

Bonnet's observations and his theory were both anticipated by Leonardo da Vinci (1452–1519). Da Vinci said in his notebook (MacCurdy, 1955) "Nature has arranged the leaves of the latest branches of many plants so that the sixth is always above the first, and so it follows in succession if the rule is not impeded." This is precisely the arrangement that Bonnet called the quincunx. Da Vinci said it is useful to plants in that "when these branches grow in the succeeding year one will not cover the other, because the five branches come forth turned in five different directions and the sixth comes forth above the first but at a sufficient distance".

SCHIMPER AND BRAUN

The foundations of the mathematical theory of phyllotaxis were laid by Karl Friedrich Schimper (1830). The theory was applied by his friend Alexander Braun (1831) in the study of the pine cone. Braun (1835) also gave an exposition of the theory in a series of articles. Schimper introduced the concepts of fundamental spiral, divergence, and parastichy. He assumed that all divergences are rational. He discovered that the most common divergences are ratios of alternate terms of the Fibonacci sequence, and that they and most of the less common divergences are convergents of the continued fraction (8). He explained the occurrence of the common divergences as the result of a fusion of two opposed tendencies in the growth of plants. One tendency was for leaves to grow as far away from each other as possible. The other was the necessity of growing nearer to each other (Schimper, 1830). Braun considered that he had explained why nature preferred Fibonacci ratios as divergences when he pointed out that they are related to the simplest of the continued fractions of the form (8).

THE BRAVAIS BROTHERS

A mathematical theory not much different from Schimper's was developed independently by L. and A. Bravais (1837). However, they rejected the idea that the convergents of equation (8) were independent rational divergences. They asserted that only the irrational numbers to which (8) converges are actual divergences, and that the rational ratios usually observed are only approximations to these. The concept of visible opposed parastichy triangle (defined in section 3) is implicit in the work of the Bravais brothers, but they never define it explicitly, nor do they ever determine a necessary and sufficient condition for the visibility of an opposed parastichy triangle.

SCHWENDENER AND VAN ITERSON

After the descriptive and mathematical work of Schimper and Braun and the Bravais brothers, attention shifted to trying to find a biological explanation for the fact that nature preferred a spiral arrangement of leaves and chose only certain divergences although others were also theoretically possible. Schwendener (1878) assumed the existence of the spirals and tried to give a mechanical causal explanation of the divergences. He tried to prove that the pressure of the growing leaves against each other would compel the divergence to shift toward the particular values that are observed in plants. His proof failed, however, because he unconsciously assumed what he was trying to prove when he assumed without proof that the successive opposed parastichy pairs that are conspicuous are (F_n, F_{n-1}) with n increasing with time (see end of section 6). Van Iterson (1907) made another attempt that failed for the same reason. A proof of Schwendener's conjecture that *under certain conditions* contact pressure pushes the divergence ever closer to $(t + \tau^{-1})^{-1}$ is given in section 11.

WIESNER AND WRIGHT

Wiesner (1875) returned to the functional explanation of da Vinci and Bonnet that the openness of a spiral arrangement of leaves gives them access to what they need for growth. The need that Wiesner stressed was the need for light, and he claimed to show experimentally (Wiesner, 1875, 1902, 1903) that a divergence of τ^{-2} caused the least shading of lower leaves by upper leaves and thus maximized the light that they received. However, the experimental data he reported do not support this claim. Wright (1873) favored the same functional explanation, and drew diagrams that showed that of all the possible rational divergences whose denominators (when the fractions are reduced to lowest terms) are less than 14, those that are closest to τ^{-2} give the most open arrangement of leaves. Wright recognized that divergences equal to the convergents of $(t + \tau^{-1})^{-1}$ have the property given by Corollary 3 of Theorem 7 (section 8). He attributed the utility of these numbers as divergences to their possession of this property.

AIRY

Airy (1873) also gave a functional explanation, but related it to the compact packing of the embryo leaves in the bud rather than to the openness of the arrangement of the adult leaves on a stem. He said that the most common divergences permitted the densest packing of the embryo leaves so that "the bud is enabled to retire into itself and present the least surface to outward danger and vicissitudes of temperature".

DeCANDOLLE

DeCandolle (1881) sought a mathematical basis for the theories of Wiesner and Airy by looking for a property of the irrational limits of continued fraction (8) that would distinguish them from all other possible divergences.

He found such a property in a geometric consequence of the fact that the continued fractions (8) have no intermediate convergents.

SACHS

Julius von Sachs, whose books (1882, 1887, 1906) on botany were standard reference works for decades, rejected the entire mathematical theory of phyllotaxis as "a sort of geometrical and arithmetical playing with ideas" (1887). He said of the Fibonacci ratios in the Schimper–Braun theory, "These fractions (parts of a continued fraction) appeared to constitute the expression of a mysterious law which was assumed to dominate growth in a supposed spiral manner. But it was nevertheless seen to be necessary, in addition to the relations of phyllotaxis represented by that fraction, to add yet others, which of course led to any continued fraction whatever, whereby however the point lost in significance". Sachs came to this incorrect conclusion because he mistakenly thought that continued fraction (8) gives "any continued fraction whatever".

TAIT AND THOMPSON

Tait (1872) wrote his note "On Phyllotaxis" after seeing but not reading through the paper of the Bravais brothers. He outlined a method of using parastichy numbers to identify the fundamental spiral and calculate its divergence. He indicated that if m and n are relatively prime and are the parastichy numbers of two intersecting sets of parastichies, with $m > n$, then there are also present two intersecting sets whose parastichy numbers are $m-n$ and n. Thus, he argued, parastichy numbers lead to the recurrence relation (1), and this inevitably leads to a Fibonacci sequence. He concluded from this argument that in the case where the divergence is between 1/3 and 1/2 the parastichy numbers "for the most conspicuous spirals must be of the forms 2, 3, 5, 8, etc., and 1, 2, 3, 5, etc." Tait was using, in effect, the concept of contraction of a visible opposed parastichy pair defined in section 6 of this paper, and the concept of extension, which is the reverse of contraction. However, his argument failed, and led him to a false conclusion, because he assumed incorrectly that a contraction is uniquely reversible, and he failed to determine the necessary and sufficient condition for visibility of an extension of a visible opposed parastichy pair (see Theorem 1 in Appendix B).

Thompson (1942) devoted Chapter XIV of his book *On Growth and Form* to the subject of phyllotaxis. His entire discussion was based on Tait's note, including Tait's conclusion quoted above. He added to Tait's conclusion these conclusions of his own: The precise value of the divergence of the fundamental spiral does not affect the parastichy numbers of the intersecting parastichies that may be visible, so it is "essentially unimportant". The fact that the rational divergences of Schimper represent a convergent series whose limit is τ^{-2} "is seen to be a mathematical coincidence, devoid of biological significance". Theorem 2 shows that the first conclusion is wrong, and Theorem 8 shows that the second conclusion is wrong.

CHURCH

Church (1904, 1920) looked for the cause of the phenomena of phyllotaxis in what happens at the growing tip of a stem. He considered this sufficient reason to reject as totally wrong any representation of leaves as points on a cylinder. He substitued for the cylindrical picture the representation of leaves as points inside a disc. He rejected the idea of a fundamental spiral, and insisted instead that the parastichies are fundamental. He put forward the theory that impulses of energy travel away from the center of the disc in spiral paths, and that new leaves grow where the spirals intersect. By rejecting the simple cylindrical representation in favor of the disc (centric) representation, Church transformed simple geometrical relationships into seemingly complicated ones, and thus made them more difficult to discover. As a matter of fact, the two representations are mathematically equivalent via transformation (12) in section 3. By rejecting the fundamental spiral in favor of parastichies, he blinded himself to the fact that the existence of each implies the existence of the other (see section 6). Because of his confusions and errors, Church led three generations of sudents of phyllotaxis into a blind alley.

SCHOUTE

Schoute (1913) argued that although contact pressure might explain how the divergence changes in a leaf distribution that is already regular, it cannot explain how leaf primordia come to have a regular distribution in the first place. To fill this gap in the theory he advanced the hypothesis that each leaf primordium secretes an inhibitor that prevents another primordium from emerging too close, and that the location of a new primordium is determined by the inhibition emanating from the two nearest primordia that are already growing. Recent students of phyllotaxis feel that the Schoute hypothesis is supported by the results of experiments in which new primordia were isolated surgically from their neighbors (Wardlaw, 1968). What is required to develop a field theory based on the principal ingredient of Schoute's hypothesis is discussed in section 16.

COXETER

H. S. M. Coxeter (1961) devotes Chapter 11 of his book *Introduction to Geometry* to the golden section and phyllotaxis. In his discussion of the pineapple, he showed that we can account for the observed parastichy numbers by assuming, as do the Bravais brothers, that the divergence is τ^{-2}. He also showed that the number of parastichies that would be most conspicuous depended on the rise, and he derived a formula that expressed this dependence. Coxeter (1972) had also rediscovered independently the connection that DeCandolle had found between a geometric property of the divergence τ^{-2} and the absence of intermediate convergents in its representation as a continued fraction.

TURING

Turing tried to develop a theory of cylindrical lattices that he could combine with his earlier paper on a chemical theory of morphogenesis to produce a theory of phyllotaxis. His work in this direction was unfinished at the time of his death in 1954. Notes on Turing's lectures on phyllotaxis taken by Hoskin and Richards are being prepared for inclusion in his collected works (Turing, A. M., unpublished, Turing, S., 1959). His theory of cylindrical lattices, as revealed by these notes, did not get very far. His concept of "principal vectors" (vectors from a lattice point or leaf primordium to its nearest neighbors) focussed his attention immediately on the most conspicuous parastichies, and thus prevented him from seeing and using the more fundamental concept of *visible parastichies* that is needed to determine just how parastichy numbers are related to the divergence of a fundamental spiral. He also never brought into his theory the number of leaves and the magnitude of the diameter of the leaf base, two of the variables which are significant in determining the changes of the divergence under contact pressure. (See section 8 to 12.)

SNOW

Mary and Robert Snow initiated an experimental phase of the study of phyllotaxis, in which they showed that the phyllotaxis of a growing plant can be altered by surgical intervention. Although Richards (1948, 1951) and Wardlaw (1968) found support for Schoute's inhibitor theory in the results of these experiments, the Snows themselves interpreted these results differently. They favored a space-filling theory that "each new leaf is determined in the first space on the growing apical cone that attains a necessary minimum size and minimum distance below the extreme tip". (Snow, M. & R., 1962).

F. J. RICHARDS

Richards, following Church, displayed a leaf distribution as a set of points in a disc. He introduced the term "plastochrone ratio" for the ratio R of the distances of two successive leaves from the center of the disc, and derived a formula relating this ratio to the number of parastichies that are most conspicuous (Richards, 1948). For the case of orthogonal parastichies, Richards' formula is equivalent to Coxeter's (The proof of this equivalence is given in Appendix C.) Later he introduced the concept of "phyllotaxis index", defined as $0 \cdot 38 - 2 \cdot 39 \log_{10} \log_{10} R$. The constants in the definition are so chosen that successive integral values of the index correspond to successive orthogonal pairs of conspicuous parastichies (Richards, 1951). Richards favored an explanation of phyllotaxis based on a "field theory" which assumes that each leaf primordium exerts an inhibiting influence on the growth of others, and that this influence diminishes with distance. However, he never developed any field equations, without which a "field theory" exists only in name. His belief that his field theory successfully explained Fibonacci phyllotaxis was not justified (see section 15).

J. theor. Biol. (1975) **53**, 435–444

A Model of Space Filling in Phyllotaxis

Irving Adler

North Bennington, Vermont 05257, *U.S.A.*

(*Received* 21 *October* 1974)

A mathematical model of the Snow space-filling theory of phyllotaxis is constructed. It is shown that a space-filling theory alone can explain why a leaf distribution tends toward equal spacing on a helix, but cannot explain why the divergence tends toward the ideal angle. A Schwabe interpretation of the Richards theory of phyllotaxis is shown to be a special case of the space-filling theory which, while it cannot account for the ideal angle, can account for the occurrence of (2, 3) and (3, 5) phyllotaxis.

1. Purpose

The space-filling theory of phyllotaxis consists of an empirical rule of Hofmeister combined with a hypothesis supplied by Snow & Snow (1962). The Hofmeister rule "that each new leaf arises in the largest gap or depression between the existing leaves or other members surrounding the apex" (Snow, 1955) indicates *where* the new leaf arises. The Snow hypothesis that ". . . each new leaf is determined in the first space on the growing apical cone that attains a necessary minimum size and minimum distance below the extreme tip" (Snow & Snow, 1962) indicates *when* the new leaf arises, namely, when the stated size and distance requirement is met. The Snows, while favoring the space-filling theory, were aware that ". . . the serious difficulty has remained that it has not been clear how a theory of this kind can account for the *exact* regulation of phyllotaxis systems to their usual fairly high accuracy . . ." (Snow & Snow, 1962).

In this paper we examine the question of just how much of the phenomena of phyllotaxis *can* be accounted for by a space-filling theory alone. To do so, we construct a mathematical model of space-filling in which the Snow terms "space" and "size" are given simple meanings that appear reasonable and are precise enough to provide an unambiguous answer to the question. The answer found is as follows. (1) The space-filling model can account for the fact that successive leaves lie on a genetic spiral, spiralling steadily to the right or to the left instead of zig-zagging at random. (2) The space-filling model can account for the fact that successive divergences of the leaves on the genetic spiral tend toward a limit with a value d between 0 and $\frac{1}{2}$. (3) The space-filling model cannot account for d having the value that corresponds to the ideal angle of normal phyllotaxis, namely, $d = \tau^{-2}$, where τ = the golden section = $\frac{1}{2}(1+\sqrt{5})$, because this requires the unlikely *ad hoc* assumption that the parameter on which the value of d depends must have the value τ^{-1}. (4) However, a special case of the space-filling model, in which this parameter lies within a particular reasonably wide interval, can account for the occurrence of (2, 3) and (3, 5) phyllotaxis. This fact is important, because (2, 3) and (3, 5) phyllotaxis are the most common cases that occur (Snow, 1955). (5) An interpretation by Schwabe of the Richards theory of phyllotaxis tranforms it into precisely this special case of the space-filling model.

2. Conventions and Notation

Let s be a positive number such that if c is a circle on the apical cone whose geodesic distance from the extreme tip is s, then a new leaf primordium is never produced above c (see Fig. 1).

FIG. 1

There are many possible choices for the distance s and the circle c associated with it. At the end of section 5 we shall show that, although the circle c is not unique, the model we construct with its help is well-defined.

By an appropriate transformation the distribution of leaf centers on the apical cone can be represented on a cylindrical surface, normalized by taking the girth of the cylinder as unit of length (Adler, 1974). By this transformation the circle c on the apical cone is transformed into a circle on the cylindrical surface. For the sake of simplicity of notation we shall label this circle c, too.

For $i > 0$, leaf i can be joined to leaf $i-1$ by two geodesic paths on the cylindrical surface. Assume the two paths are unequal in length. If the shorter path from leaf $i-1$ to leaf i goes up to the right, we say that leaf i is to the right of leaf $i-1$; otherwise that leaf i is to the left of leaf $i-1$. Without loss of generality we assume that leaf 1 is to the right of leaf 0.

For $i > 0$ denote by $d_i > 0$ the fraction of a turn around the axis of the cylinder made when passing from leaf $i-1$ to leaf i by the shorter path between them. Then $d_i < \frac{1}{2}$. Denote by r_i the length of the component of this path that is parallel to the axis of the cylinder. r_i is the normalized internode distance between leaf i and leaf $i-1$.

Let leaf $n-2$ and leaf $n-1$ be the last two leaves that have emerged. Slit the cylindrical surface along the element l through leaf $n-2$, and unroll the surface on a plane. Then the circle c becomes a segment of a straight line perpendicular to the element through leaf $n-2$. Let f be the line through leaf $n-1$ parallel to c. Let m be the element through leaf $n-1$. The lines c, f, l and m determine two rectangles $R_1(n)$ and $R_2(n)$, on the left and right respectively, as shown in Fig. 2. Denote their height by h_n. If leaf $n-1$ is to the right of leaf $n-2$, then the base of $R_1(n)$ has length $d_{n-1} < \frac{1}{2}$, and the base of $R_2(n)$ has length $1-d_{n-1} > \frac{1}{2}$. Then the area of $R_2(n)$ is greater than the area of $R_1(n)$. As the apical cone grows, the line c moves up away from leaf $n-1$, so that the height h_n increases.

3. Assumptions

We are now ready to make the assumptions which define the space-filling model.

Assumption 1

If leaf $n-1$ is to the right of leaf $n-2$, leaf n emerges with its center in the interior or on the upper base of the larger of the two rectangles $R_1(n)$ and $R_2(n)$ at the moment when that rectangle attains the area $k > 0$.

Since leaf 1 is to the right of leaf 0, $R_2(2)$ is larger than $R_1(2)$, hence leaf 2 emerges with its center in the interior or on the upper base of $R_2(2)$.

We now seek the weakest assumption that will assure that when leaf 2 emerges it will be to the right of leaf 1.

Let s_2 be the geodesic path from leaf 1 to leaf 2 going up to the right. (See Figure 2 with n taken equal to 2.) Denote by p_2 the horizontal component of s_2. The vertical component of s_2 is r_2. Then $p_2 = a_2(1-d_1)$, $0 < a_2 < 1$; and $r_2 = b_2 h_2$, $0 < b_2 \leqslant 1$. Leaf 2 is to the right of leaf 1 if and only if $p_2 < \frac{1}{2}$. Suppose that

$$0 < x < d_1 < \frac{1}{2}. \tag{1}$$

Then $d_2 = p_2 < \frac{1}{2}$ if $a_2 \leqslant \frac{1}{2}$, no matter what the value of x may be. However, if $a_2 > \frac{1}{2}$, the permitted values of a_2 that yield $d_2 = p_2 < \frac{1}{2}$ are linked to the value of x. Let us require not only that $d_2 = p_2 < \frac{1}{2}$, but

$$x < d_2 < \frac{1}{2},$$

84

438   I. ADLER

so that condition (1) is reproduced for the next higher index of d_n. This requirement is met if we restrict a_2 to the range

$$2x < a_2 < 1/2(1-x). \qquad (2)$$

This range is maximized while still satisfying the condition $a_2 > \frac{1}{2}$ when $x = \frac{1}{4}$, and $\frac{1}{2} < a_2 < 2/3$. Accordingly, we make Assumption 2.

ASSUMPTION 2

$$\frac{1}{4} < d_1 < \frac{1}{2}$$
$$0 < a_2 < 2/3 \qquad (3)$$

We now proceed by induction on the leaf number n. If leaf $n-1$ is to the right of leaf $n-2$, $R_2(n)$ is larger than $R_1(n)$, and leaf n emerges in the interior or on the upper base of $R_2(n)$. We define s_n, p_n and r_n inductively as shown in Fig. 2 and described explicitly below.

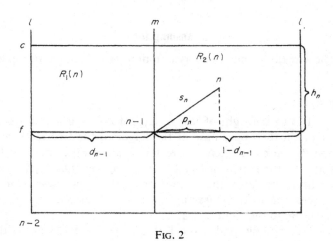

FIG. 2

Let s_n be the geodesic path from leaf $n-1$ to leaf n going up to the right. Denote by p_n the horizontal component of s_n. The vertical component of s_n is r_n. Denote $p_n/(1-d_{n-1})$ by a_n, and r_n/h_n by b_n. To simplify the model, we make Assumption 3.

ASSUMPTION 3

$$a_n = a, \quad \text{for all } n \geqslant 2.$$

Then

$$p_n = a(1 - d_{n-1}), \quad r_n = b_n h_n, \quad 0 < a < 2/3, \quad 0 < b_n \leqslant 1. \tag{4}$$

4. The Genetic Spiral

PROPOSITION 1

In the model defined by Assumptions 1, 2 and 3, leaf n is to the right of leaf $n-1$ for every $n \geqslant 1$. Hence successive leaves lie on a genetic spiral that spirals up to the right.

The proof is by induction on n carried out separately for the two cases $a \leqslant \frac{1}{2}$ and $\frac{1}{2} < a < 2/3$. In the case $a \leqslant \frac{1}{2}$, we have $\frac{1}{4} < d_1 < \frac{1}{2}$ by Assumption 2, and whenever $d_{n-1} < \frac{1}{2}$,

$$d_n = a(1 - d_{n-1}), \tag{5}$$

and

$$d_n < \tfrac{1}{2}.$$

In the case $a > \frac{1}{2}$, we have $\frac{1}{4} < d_1 < \frac{1}{2}$, and $\frac{1}{2} < a < 2/3$, and whenever

$$\tfrac{1}{4} < d_{n-1} < \tfrac{1}{2}, \tag{6}$$

d_n is defined by (5), and (6) holds with $n-1$ replaced by n.

5. Convergence of d_n

From the recurrence relation (5) we get

$$d_n = a - a^2 + \ldots + (-1)^n a^{n-1} + (-1)^{n-1} a^{n-1} d_1. \tag{7}$$

Then

$$d_n = (a + (-1)^n a^n)/(a+1) + (-1)^{n-1} a^{n-1} d_1. \tag{8}$$

Then, since

$$0 < a < \tfrac{2}{3}, \, d = \lim_{n \to \infty} d_n$$

exists, and

$$d = a/(a+1). \tag{9}$$

From Assumption 1,

$$h_n = k/(1 - d_{n-1}), \quad \text{for } n \geqslant 2. \tag{10}$$

Since d_n converges, so does h_n. Then the rectangles $R_2(n)$ tend toward congruence as n increases. It is reasonable then to make Assumption 4.

ASSUMPTION 4

b_n converges to a limit b as $n \to \infty$.
In view of (4),

$$r_n = b_n k/(1-d_{n-1}),\tag{11}$$

and

$$r = \lim_{n \to \infty} r_n = bk/(1-d) = bk(a+1).\tag{12}$$

Therefore we have Proposition 2.

PROPOSITION 2

The genetic spiral tends to become helical, and the intervals between successive leaves on the genetic spiral tend toward equality as n increases.

Possible choices for the circle c

We can now identify all the circles that satisfy the condition used in section 2 to define c. There is associated with circle c an area k, with k a function of c such that a new leaf emerges with center under or on c when the area of $R_2(n) = k$. The distance from the center of the new leaf to c is

$$h_n - r_n = (1-b_n)k/(1-d_{n-1}).$$

The sequence of numbers $(h_n - r_n)$ has a greatest lower bound $e \geqslant 0$. Let C be the circle parallel to c and below c in the cylindrical representation at a distance equal to e. This circle is uniquely defined. It is that one of all possible circles fitting the definition of c that is furthest from the tip of the stem. Every circle above and parallel to C fits the definition of c.

With each choice of c there are associated the parameters a, b_n, k, h_n by which we determined the successive values of d_n and r_n, and, ultimately, d and r. The value of a is independent of the choice of c. Since the value of a determines the values of d_n and of d, these values are also independent of the choice of c. To assure that the position of each new leaf primordium is independent of the choice of c, we must specify that r_n is independent of the choice of c. Then it follows from equation (11) that since the value of $1-d_{n-1}$ is independent of the choice of c, so is the value of $b_n k$. Thus, if c is varied by moving it up toward the tip of the stem from C, k increases, h_n for each n varies as k, and b_n for each n varies inversely as k. Consequently, once the values of the parameters are fixed for a particular choice of c, the values of the corresponding parameters are determined for every choice of c, and the model is well-defined.

The numbers k and s represent the "necessary minimum size" and "minimum distance below the extreme tip" referred to by Snow & Snow (1962) in their hypothesis. Although these numbers differ for different choices of c, the possibility of choice does not introduce any ambiguity into the model.

If we do not require that $a_n = a$ for all $n \geqslant 2$, the recurrence relation (5) is replaced by

$$d_n = a_n(1 - d_{n-1}).\qquad(13)$$

Then (7) is replaced by

$$d_n = a_n - a_n a_{n-1} + \ldots + (-1)^n \prod_{i=2}^{n} a_i + (-1)^{n-1} d_1 \prod_{i=2}^{n} a_i.\qquad(14)$$

Since (7) converges, then, for values of n that are sufficiently large, and any positive integer j, successive values $d_n, d_{n+1}, \ldots, d_{n+j}$ given by (7) are approximately equal. Since (14) defines a continuous function of $a_1, \ldots, a_n$ for every value of n, successive values of $d_n, \ldots, d_{n+j}$ given by (14) for sufficiently large n are approximately equal to those given by (7) if $a_2, \ldots, a_n$ are close enough to a. Hence the model is capable of explaining why a leaf distribution tends toward equal spacing on a helix as n increases even if we do not require strict equality of successive values of a_n, but merely require that for large enough values of n successive values of a_n are approximately equal.

6. The Ideal Angle

Since $0 < a < 2/3$, we find from equation (9) that

$$0 < d < 2/5,\qquad(15)$$

and, as long as no further restriction is placed on the values of a that are permitted, every value of d given by (15) is possible. On the other hand, if $d = \tau^{-2} = $ the ideal angle, then from equation (9)

$$a/(a+1) = \tau^{-2},\qquad(16)$$

and, from the property of τ that

$$\tau^2 = \tau + 1,\qquad(17)$$

it follows that

$$a = \tau^{-1}.\qquad(18)$$

That is, for the space-filling model to assure convergence of d_n to the ideal angle, it is necessary to make the *ad hoc* assumption that the projection of the center of each new leaf on the base of the rectangle in which it arises divides that base in such a way that the ratio of the base to the larger of the two segments into which it is divided is the golden section. A model using such a strong assumption explains nothing, because it assumes rather than accounts for the "exact regulation of phyllotaxis systems". Hence we have Proposition 3.

PROPOSITION 3

The space-filling model defined by Assumptions 1, 2, 3 and 4 without the *ad hoc* assumption that $a = \tau^{-1}$ cannot explain convergence of d_n to τ^{-2}.

7. A Space-filling Interpretation of the Richards Theory of Phyllotaxis

Richards (1948, 1951) proposed a field theory of phyllotaxis based on the suggestion by Schoute (1913) that the stem apex and each leaf primordium secrete an inhibitor that prevents a new leaf from emerging too close to them. Adler (1974) listed three assumptions which are implicit in Richards' discussions of a field theory and designated these as assumptions R. The Richards argument was in essence that assumptions R implied (i) that while the distance from the stem apex at which a new leaf primordium arises is determined chiefly by the apex, its position at that level is determined principally by the two primordia that are its nearest neighbors, and that the primordium is formed "tangentially somewhat nearer the older one" of these two determining primordia; and that (i) in turn implied (ii) that successive d_n converge to τ^{-2}, the ideal angle of normal Fibonacci phyllotaxis. Consequently the Richard theory culminates in the assertion that assumptions R imply (ii). Adler (1974) refuted this assertion by producing a counter example.

The Richards argument contains two separate steps: assumptions R imply (i) and (i) implies (ii). If we omit the first step, we still have a theory of phyllotaxis that is worth examining in its own right, namely that (i) implies (ii). The Richards formulation of (i) contains the ambiguous term "nearest neighbors". There are several possible interpretations of this term. One of them is that the nearest neighbors are those that are nearest "tangentially", that is, of all the elements of the cylinder on which leaf primordia lie, the "nearest neighbors" lie on elements nearest to the new primordium one on each side of it. This interpretation was discussed at length in Adler (1974). Another possible interpretation is that the nearest neighbors are those that are nearest vertically, that is, that are on the first two levels below the new leaf primordium where older leaf primordia are found. In this interpretation, the "nearest neighbors" of a new leaf are its two immediate predecessors. This interpretation of the Richards formulation was given to it by Schwabe (1972): ". . . the degree of accuracy in the spacing required can be fully satisfied if the new organ falls into the larger gap between its two predecessors and somewhat nearer to the older of the two."

With this Schwabe interpretation, the Richards theory that (i) implies

(ii) does not differ in essence from a space-filling theory, since the new organ is formed in the rectangle above "the larger gap" when that rectangle is large enough to contain a point where the magnitude of the inhibition has become small enough to be ineffective. Therefore the space-filling model developed here can be used to test the Richards–Schwabe theory that (i) implies (ii). The assumption in the theory that a new primordium is "somewhat nearer to the older of the two" primordia that determine its position is equivalent to specifying that $a > \frac{1}{2}$. Hence the Schwabe interpretation of the Richards theory can be tested by using a model based on Assumptions 1, 2, 3 and 4 and the condition

$$\frac{1}{2} < a < 2/3. \tag{19}$$

Condition (19) restricts the divergence d to the interval

$$1/3 < d < 2/5 \tag{20}$$

but permits d to take on any value in this interval. Therefore the Richards–Schwabe theory cannot account for d being restricted to the value τ^{-2} of the ideal angle. However, as shown in Adler (1974), the condition (20) implies that the opposed parastichy pairs (2, 3) and (5, 3) are both visible, and that therefore the phyllotaxis could be (2, 3) or (3, 5) if r, the normalized internode distance, has the appropriate value. Moreover, (2, 3) and (3, 5) phyllotaxis are the most common examples of phyllotaxis (Snow, 1955). Therefore, while the Richards–Schwabe theory does not suffice to explain the genesis of the "ideal angle" in phyllotaxis, it does suffice to explain the genesis of the divergences found in the most common examples of phyllotaxis. Hence we have Proposition 4.

PROPOSITION 4

The Schwabe interpretation of Richards' theory is equivalent to the space-filling model with $a > \frac{1}{2}$. It cannot explain the convergence of d_n to τ^{-2}, but it can explain the occurrence of divergences that produce (2, 3) and (3, 5) phyllotaxis.

8. Foundations for a Complete Model of Fibonacci Phyllotaxis

In Adler (1974) it was shown that contact pressure could explain the convergence of d to τ^{-2} if these three assumptions are made: (1) that successive divergences are equal to some value d; (2) that $1 \cdot 14 \leqslant T^2 r(T) < 1 \cdot 79$, where T is the time measured in plastochrones from the emergence of leaf 0, and $r(T)$ is the value of r at time T; (3) that contact pressure, defined as maximization of the minimum distance between leaves, begins early (when

$T < 5$). In new results now being prepared for publication, it is shown that the conclusion still holds if assumption (2) is replaced by the more general assumption that r is a monotonic decreasing function of T. In Adler (1974) it was shown, too, that an inhibitor field theory based on assumptions R could account for divergences tending toward equality so that a field theory combined with contact pressure could be the basis for a complete model of Fibonacci phyllotaxis.

We have shown above that a space-filling theory characterized by Assumptions 1, 2, 3 and 4 can also account for divergences tending toward equality. Thus a space filling theory combined with contact pressure can also be the basis for a complete model of Fibonacci phyllotaxis. Consequently, in so far as utility for model building is concerned, neither of the two rival theories of the 1940s [the space-filling theory of Snow & Snow (1962) and the inhibitor theory of Richards] has any advantage over the other. Neither alone can explain the tendency of d to converge to τ^{-2}, but either one can do so when combined with the contact pressure model, if contact pressure is assumed to begin early and if r is assumed to be a monotonic decreasing function of time.

I am grateful to Peggy Adler for the figures.

REFERENCES

ADLER, I. (1974). *J. theor. Biol.* **45,** 1.
RICHARDS, F. J. (1948). *Symp. Soc. exp. Biol.* **2,** 217.
RICHARDS, F. J. (1951). *Phil. Trans. R. Soc. B* **235,** 509.
SCHOUTE, J. C. (1913). *Recl. Trav. bot. néerl.* **10,** 153.
SCHWABE, W. W. (1972). Personal communication.
SNOW, M. & SNOW, R. (1962). *Phil. Trans. R. Soc. B* **244,** 483.
SNOW, R. (1955). *Endeavour* **14,** 190.

J. theor. Biol. (1977) **65,** 29–77

The Consequences of Contact Pressure in Phyllotaxis

Irving Adler

North Bennington, Vermont 05257, *U.S.A.*

(*Received 16 December* 1975)

To determine the consequences of contact pressure in phyllotaxis, a mathematical model is constructed in which a leaf distribution is represented by a point lattice of $n + 1$ lattice points at equal intervals on a helix wound around a cylinder. The model is normalized by taking the girth of the cylinder as 1 and by measuring time T in plastochrones, so that $n = [T]$. r stands for the normalized internode distance (component of the distance between two consecutive lattice points that is parallel to the axis of the cylinder). d stands for the divergence (fraction of a turn between consecutive lattice points). It is assumed that r is a monotonic decreasing function of T such that $r(T) \to 0$ as $T \to \infty$. Contact pressure is represented by the assumption that the minimum geodesic distance between lattice points is maximized. It is shown that if (p, q), with $p < q$, is the contact phyllotaxis determined when contact pressure first becomes effective, then the continuation of contact pressure requires that the advance to higher phyllotaxis as r decreases must proceed via successive pairs of consecutive terms of the Fibonacci sequence generated by the numbers p and q, namely, $p, q, p + q, p + 2q, 2p + 3q, \ldots$. The divergence, starting from some value $d = 1/t + 1/a_2 + \ldots + 1/(a_n + x)$† determined by p and q converges to an ideal angle $1/t + 1/a_2 + \ldots + 1/a_n + 1/\tau$, where τ is the golden section. A necessary and sufficient condition for the ideal angle to be $1/2 + 1/\tau = \tau^{-2}$ is that the p and q of the initial contact phyllotaxis be consecutive Fibonacci numbers of the sequence 1, 2, 3, 5, 8, $\ldots$. It is proved that a sufficient condition for convergence to the ideal angle τ^{-2} of normal phyllotaxis is that contact pressure begin before $T = 5$ or before $r < 3^{\frac{1}{2}}/38$ with d initially between 1/3 and 1/2.

1. The Problem, Procedures and Assumptions

(1) This paper relates to the problem of explaining why the numbers of conspicuous left spirals and right spirals in a leaf distribution on a plant are usually consecutive Fibonacci numbers (normal phyllotaxis).

(2) Adler (1974) gave an existence proof showing that there exist conditions under which contact pressure would suffice to produce normal

---

† This is a continued fraction written in the convention employed throughout this paper: everything written after a / is understood to be under the fraction line it represents.

29

phyllotaxis. This paper answers the question, "what are the consequences of contact pressure under all possible conditions?" As corollaries of the answer we obtain a necessary and sufficient condition for contact pressure to produce normal phyllotaxis, and a very wide sufficient condition to do so.

(3) The investigation requires the use of three different surfaces: (a) the surface on which the leaf distribution exists; (b) a cylinder on which the leaf distribution is represented; (c) a phase space for the latter.

(4) The surface on which a leaf distribution is found may be a cylinder, a disc, a cone, a paraboloid, or generally any surface of revolution.

(5) We may, without loss of generality, consider only the case of one fundamental spiral going up to the right (Adler, 1974).

(6) *Definitions.* The term *leaf* is used here to refer to the region on the surface of a stem where a leaf or floret or scale is attached. Usually it will suffice to represent a leaf by its center. A *plastochrone* is the interval of time between the emergence of two successive leaves. The *fundamental spiral* is obtained by joining each leaf to the next higher one (next younger one) by the shortest path on the surface. *Parastichies* are secondary spirals determined by joining a leaf to any other leaf. Parastichies occur in parallel sets, some going up to the left and some going up to the right. The *conspicuous parastichies* are those determined by joining each leaf to its nearest neighbors to the left and right. In *normal phyllotaxis* the two numbers obtained by counting the conspicuous left parastichies and the conspicuous right parastichies are consecutive terms of the sequence $1, t, t+1, 2t+1, \ldots$ in which each term after the first two is the sum of the two terms immediately preceding it. In *anomalous phyllotaxis* they are consecutive terms of a sequence whose first two terms are t and $2t+1$ respectively, with later terms generated by the same addition rule as in normal phyllotaxis. We shall be concerned principally with the case $t = 2$.

(7) By means of an appropriate transformation we can get a cylindrical representation for any leaf distribution. Disc → cylinder (Adler, 1974). Cone → disc (Richards, 1951). For an arbitrary surface of revolution, divide the surface into the narrow parallel zones between consecutive leaves. Each zone is approximately a zone of a cone. Then transform: cone → disc → cylinder. The details of these transformations are given in Appendices A and B.

(8) The cylindrical representation is normalized by taking the girth as unit of length and the plastochrone as unit of time T. Leaves are numbered $0, 1, 2, \ldots$. The state of a system is determined by d_i and r_i, the horizontal (rotational) and vertical components respectively of the geodesic distance between leaves i and $i-1$, $i = 1, 2, 3, \ldots$. d_i and r_i are functions $d_i(T)$ and $r_i(T)$ of time. The leaves present at time T are $0, 1, \ldots, [T]$.

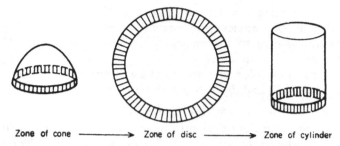

Zone of cone ⟶ Zone of disc ⟶ Zone of cylinder

FIG. 1. Cylindrical representation of a zone on a surface of revolution via the transformation given in Appendices A and B.

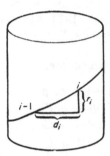

FIG. 2. Normalized cylindrical representation.

The following additional symbols will be used:

l_0 = the element of the cylinder through leaf 0.
dist $(0, n)$ = the geodesic distance on the cylinder from leaf 0 to leaf n.
dist (l_0, n) = the distance on the cylindrical surface from leaf n to l_0.
min dist^2 $(0, n)$ = the square of the minimum of dist $(0, n)$ for $n \le [T]$.

(9) *Assumptions.* (1) At each instant T, the $d_i(T)$ for all $i \le [T]$ are equal, with a value $d(T)$ that depends only on T and not on i; and the $r_i(T)$ for all $i \le [T]$ are equal, with a value $r(T)$ that depends only on T and not on i. Therefore, at each instant the state of a leaf distribution is represented by a point (d, r) in the (d, r) plane (the phase space), or, equivalently, by the position of leaf 1 relative to horizontal and vertical axes through leaf 0 in a plane development of the cylindrical representation. By definition, $d \le 0.5$. We shall assume that $r < 3^{\frac{1}{2}}/6$. (See note 4 below.)

(2) $r(T)$ is a monotonic decreasing function of T, with

$$\lim_{T \to \infty} r(T) = 0.$$

(3) Contact pressure assumption: For $T \geq T_c$, $d(T)$ is such that the minimum distance between leaves in the normalized cylindrical representation is a maximum relative to the values of the minimum distance that correspond to neighboring values of d.

Note 1: see section 3 for a discussion of assumption (3).

Note 2: when $r_i(T)$ and $d_i(T)$ vary with i, we may use

$$r = \bar{r}_i = (r_1 + \ldots + r_i)/i, \qquad d = \bar{d}_i = (d_1 + \ldots + d_i)/i.$$

Note 3: in this paper we deal principally with divergences in the range $1/3 < d < 1/2$, which is the case $t = 2$ of $1/(t+1) < d < 1/t$. The results are easily extended to the other cases $t = 3, 4, \ldots$, where the extension is not already stated explicitly.

Note 4: for $r \geq 3^{\frac{1}{2}}/6$, maximization of the minimum distance between leaves requires that $d = \frac{1}{2}$ (Adler, 1974, Appendix D).

(10) *Examples.* Some functions $r(T)$ that satisfy assumption 2:

$$\left. \begin{array}{l} r = T^{-1}/4 \\ r = T^{-0.3}/15 \end{array} \right\} \text{(arbitrarily chosen).}$$

$r = 1 \cdot 5T^{-2}$ (used for existence proof in Adler, 1974).

$r = (2\pi)^{-1} \ln \{(1 + aT)/[1 + a(T - 1)]\}, \quad a > 0.$

(Derived from a disc model.) (See Appendix A.)

$r = (4\pi)^{-1}[1 + 4a(T + b)]^{\frac{1}{2}} \ln [(T + b)/(T + b - 1)], \quad a, b > 0.$

(Derived from a parabolic model.) (See Appendix B.)

2. The Bare Apex

An important empirical fact about phyllotaxis is that there is a bare region at the top of the growing stem where no leaf primordia are ever found. (Schoute, 1913, p. 176; Richards, 1948, p. 226, 1951, p. 559). Any model of phyllotaxis must take it into account. The Schoute–Richards inhibitor theory of phyllotaxis takes it into account by postulating that the stem apex secretes an inhibitor that prevents the initiation of leaf primordia at the dome apex. However, the fact that the apex is bare seems to imply more than that primordia cannot be initiated there. It seems to imply, too, that primordia that are initiated elsewhere cannot grow into it later. That is, the center of the apex dome is a *forbidden zone*. Without being committed to any theory that may explain it (whether it be the secretion of an inhibitor at the apex, or the simple fact that the apex itself constantly grows *away* from existing primordia), we shall postulate the existence of this forbidden zone at the center of the apex dome as an empirical fact. In section 3 we shall see that this postulate plays a part in determining the content of assumption 3, the contact pressure assumption.

3. The Contact Pressure Assumption

Assumption 3 of section 1 requires an explanation. We are picturing each leaf distribution under consideration in two ways: (a) as a distribution on the surface of the stem; (b) as a normalized cylindrical representation obtained by the transformations described in Appendices A and B. The value of the divergence which maximizes the minimum distance between leaves on one of these two surfaces need not be the same as the one which maximizes the minimum distance between leaves on the other. In the formulation of assumption (3) it is therefore necessary to specify which of the two surfaces it is on which the minimum distance is being maximized.

At first sight it would seem to be natural to choose for this purpose the surface of the stem apex, and to use, instead of assumption (3), a different formulation of it as follows:

Assumption (3*): For $T \geq T_c$, $d(T)$ is such that the minimum distance between leaves on the surface of the stem apex is a maximum relative to the values of the minimum distance that correspond to neighboring values of d.

However, there are five reasons why we choose to use assumption (3) rather than assumption (3*).

(1) When many leaves are present, so that leaf 0 is far away from the center of the apex dome, the zone on the stem surface between leaf 0 and its nearest neighbor can be transformed into a cylindrical zone with only slight distortion of distances. For this reason, when many leaves are already present assumptions (3) and (3*) have essentially the same consequences. Under these conditions it doesn't make any difference which of the two assumptions is used. (This assertion has been verified by computer calculations.)

(2) However, the two assumptions have entirely different consequences if contact pressure is effective when only a few leaves are present. When only a few leaves are present, assumption (3*) requires that the regions occupied by neighboring leaves, in order to be in contact at their peripheries, must intrude into the forbidden zone at the center of the apex dome, while assumption (3) does not, because in the cylindrical representation of a leaf distribution the center of the apex dome is made inaccessible by being sent off to infinity. Consequently, when contact pressure begins early, assumption (3) is consistent with the existence of a forbidden zone on the apex dome, but assumption (3*) is not.

(3) Contact pressure does not take place only on the surface of the stem apex. It takes place in three dimensions. Although it is customary to represent a leaf primordium by the region occupied by its base on the surface of the stem, the primordium is actually a three-dimensional object growing outward

from the stem. In a narrow zone of the stem, the bases of the leaf primordia radiate from the stem like spokes of a wheel, and are analogous to a flat sheaf of lines through a point. In geometry, the metric relations in a sheaf of lines are found to be isomorphic to the metric relations of points on a circle. Analogously, it would seem to be reasonable to use relations among points on a circle to picture the contact pressure relations among leaf primordia attached to a narrow zone on the surface of the stem. Then, while the zone moves downwards to generate the whole surface of the stem, this circle moves downwards to generate a cylindrical surface. Thus the use of the cylindrical representation to picture contact pressure relations among leaf primordia is a three-dimensional analogue of the use of a circle in geometry to picture metric relations in a sheaf of lines. In any case, the cylindrical representation of a leaf distribution is not to be thought of as something that is actually seen on the plant. It is an abstraction in which a three-dimensional primordium is represented by a point on the cylinder, and the growth of each primordium and its consequent pressure on its neighbors is represented by the requirement that the minimum distance between these points be maximized.

(4) The cylindrical representation of a leaf distribution used in assumption (3) is normalized. This puts all primordia, young and old, on an equal footing and automatically compensates for inequalities of size, mutual distances, and growth rates that result from inequalities of age. This is not true of the disc picture or parabolic picture of a leaf distribution.

(5) If a contact pressure model is constructed using assumption (3*) instead of assumption (3), consequences are obtained which do not fit the facts. Specifically, it is a fact that in most plants the ". . . divergence passes rapidly and unerringly to a very close approximation to the Fibonacci angle . . ." (Richards, 1948, p. 218); and that normal phyllotaxis is often well established by the time the seventh primordium emerges (Davies, 1939). If a contact pressure model is to be consistent with this fact, then it should be capable of producing normal phyllotaxis when contact pressure begins early. However, with assumption (3*), under conditions of early contact pressure, the divergence angle takes on a value that makes the later convergence to the Fibonacci angle impossible. Computer calculations show, for example, that if initially d is between 1/3 and 1/2, and contact pressure begins at $T = 2$, d is compelled to decrease to a value below 1/3. If contact pressure begins at $T = 3$, d is compelled to take on a value above 2/5. In both cases later convergence to the Fibonacci angle becomes impossible. On the other hand, as we show in the remainder of this paper, the use of assumption (3) not only permits normal phyllotaxis when contact pressure begins early, it makes normal phyllotaxis *inevitable* under these conditions.

Reason (5) alone would suffice to justify the choice of assumption (3) instead of assumption (3*) for the construction of a contact pressure model, since the ultimate test of a mathematical model is whether or not its consequences fit the facts. The other four reasons are useful, however, in that they illuminate properties of assumption (3) that give it an advantage over assumption (3*) and help it succeed where the latter fails.

4. Leaves Nearest Zero

Let l_0 be the element through leaf 0. Let $X(nd)$ be the integer nearest nd.

$$X(nd) = [nd] \text{ or } 1 + [nd].$$
$$\text{dist}(l_0, n) = |nd - X(nd)|, \text{ and dist}^2(0, n) = n^2 r^2 + \text{dist}^2(l_0, n).$$

Hence

$$\text{dist}^2(0, n) = n^2 r^2 + [nd - X(nd)]^2. \tag{1}$$

(When the r_i are all different, nr is replaced by $r_1 + \ldots + r_n$.) Leaf a is a *leaf nearest zero* if for every leaf $n \neq 0$, dist $(0, n) \geq$ dist $(0, a)$. If $m > n$, and n is a leaf nearest zero, m can become a leaf nearest zero as r falls only if dist $(l_0, m) <$ dist (l_0, n). This leads to the concept of *points of close return*, defined inductively: $n_1 = 1$; n_{i+1} is the first leaf after n_i for which dist $(l_0, n_{i+1}) <$ dist (l_0, n_i) (Adler, 1974). Coxeter (1972) calls them *principal neighbors*. The n_i are the denominators of the successive principal convergents of the expansion of d as a simple continued fraction, with the first convergent taken to be 0/1 (Coxeter, 1972).

If the points of close return on one side of l_0 are joined in succession, a half polygon is formed. The two half polygons and l_0 are related to the famous Klein diagram by an affine transformation. For any three consecutive points of close return, n_{i-1}, n_i and n_{i+1}, $n_{i+1} = n_{i-1} + an_i$, where a is an integer ≥ 1. If $a > 1$, the intermediate leaves $n_{i-1} + n_i, \ldots, n_{i-1} + (a-1)n_i$ all lie on the straight line segment joining n_{i-1} and n_{i+1} (Coxeter, 1972).

The identities of the leaves nearest zero depend on d, which determines the n_i, on T, which determines which leaves are present, and on r, which determines which of the n_i present are nearest to leaf zero.

For given r and T, we may plot min dist$^2 (0, n)$, $n \leq [T]$, as a function of d. Figure 3 shows how min dist$^2 (0, n)$ varies with d for $r = 0 \cdot 13$, $0 \cdot 05$ and $0 \cdot 01$, with $T = 5$ and $T = 8$. For every value of $r < 3^{\frac{1}{2}}/6$ and every value of $T \geq 2$, the function has one or more maxima in the range $1/3 < d < 1/2$.

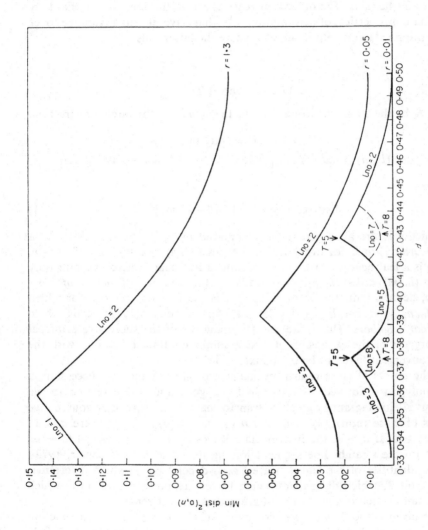

FIG. 3. min dist² $(0, n)$ as a function of d when $T = 5$ and $T = 8$. (Lno = leaf nearest zero.)

5. Maximization of min dist $(0, n)$ at $T = T_c$

The state of the phyllotaxis system has some initial value (d, r) at time T_c, the time when contact pressure first becomes effective. The values of d, r and T_c determine a value for the leaf nearest zero, say p. The invocation of assumption (3) for the first time at $T = T_c$ requires that the initial value of d be replaced by one at which dist $(0, p)$ is maximized. For the given values of r and T_c, min dist^2 $(0, n)$, which equals dist^2 $(0, p)$ at the given value of d, is a function of d, with a graph similar to those shown in Fig. 3. There are three possibilities:

(1) The initial value of d is one at which dist^2 $(0, p)$ = dist^2 $(0, g)$ is a maximum, where p and q are the leaves nearest zero for values of d near the initial value of d on opposite sides of this initial value of d. Then assumption (3) requires no change in d at $T = T_c$.

(2) The point of the graph that corresponds to the initial value of d is a minimum point. The corresponding state is unstable in a growing plant in which all parameters are constantly changing. Any small perturbation, such as that provided by the assumed decrease in r with the passage of time, will cause the point that corresponds to the state of the plant to move either to the left or to the right of the minimum point. Then case (3) will apply.

(3) The point of the graph that corresponds to the initial value of d is on a sloping section of the graph. Then d must change in that direction that permits the point $[d, \text{dist}^2 (0, p)]$ to climb the slope until it reaches the point where dist^2 $(0, n)$ is a maximum. The desired direction of change is easily determined from the formula dist^2 $(0, p) = r^2 p^2 + [pd - X(pd)]^2$. dist^2 $(0, p)$ increases as d increases if $pd > X(pd)$, while dist^2 $(0, p)$ increases as d decreases if $pd < X(pd)$. Maximization of min dist^2 $(0, n)$ will take place at some value $d = d_{max}$ whose value depends on r and T_c. Beyond this value of d, p yields its place as leaf nearest zero to another leaf q. At d_{max}, however, dist^2 $(0, p)$ = dist^2 $(0, q)$, and both p and q are leaves nearest zero. When $d = d_{max}$, an increase in d leads to an increase of one of the two distances, dist $(0, p)$ or dist $(0, q)$, while a decrease in d leads to an increase in the other. That one of the leaves nearest zero, p or q, whose distance from leaf 0 increases as d increases is necessarily to the right of l_0, while the other one must be to the left of l_0.

When $d = d_{max}$, the point (d, r) in the phase space lies on the (p, q) semicircle defined by the equation

$$\text{dist}^2 (0, p) = \text{dist}^2 (0, q), \quad r > 0. \qquad (2)$$

Then, as r decreases, assumption (3) requires that the point (d, r) descend along the (p, q) semicircle until another leaf with leaf number $s > p, q$ becomes leaf nearest zero.

We consider next two questions: (1) how are p and q related to the values of d encountered as the point (d, r) moves on the (p, q) semicircle? (2) how is s related to p and q and the value of d at which s first becomes leaf nearest zero, if s already exists when that value of d is reached?

6. How Two Leaves Nearest Zero Equidistant from Zero and on Opposite Sides of l_0 are Related to d

In the preceding section we found that when the minimum distance between leaves is maximized at $T = T_c$, two integers p and q are determined by the initial values of d and r at time $T = T_c$ such that (1) leaves p and q are both leaves nearest zero, (2) p and q are on opposite sides of l_0 and (3) dist $(0, p) =$ dist $(0, q)$. Let d be the divergence of the leaf distribution under these conditions. We already know that p and q are necessarily points of close return, and that the latter have been identified by Coxeter (1972) as demoninators of principal convergents of the simple continued fraction expansion of d. We now prove that in fact they are the denominators of *consecutive* principal convergents of the simple continued fraction expansion of d. In Adler (1974) a conspicuous opposed parastichy pair was defined as one determined by a leaf and its nearest neighbors to the right and to the left. Since p and q are both leaves nearest zero and are on opposite sides of l_0, they determine a conspicuous opposed parastichy pair. Moreover, as a consequence of assumption (3), they satisfy the additional condition that dist $(0, p) =$ dist $(0, q)$. Then, if $y =$ dist $(0, p)$ and $x =$ dist $(0, w)$, where w is any leaf other than leaf zero, then $x \geq y$. The following theorem is an immediate consequence of this fact:

Theorem 1. Under conditions of contact pressure (maximization of the minimum distance between leaves) a *conspicuous* opposed parastichy pair is *visible*.

Proof. A visible opposed parastichy pair is defined as one in which there is a leaf at every intersection of a left parastichy and a right parastichy determined by the pair. Suppose there is an intersection of a left parastichy and a right parastichy with no leaf at the intersection. Then we would have the situation shown in Fig. 4 where a, b, c and d are leaves, there is no leaf at the intersection p of ac and bd, and the lengths of ac and bd are both equal to y.

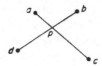

FIG. 4. Intersection of a left parastichy and a right parastichy if there were no leaf at the intersection.

There must be two of the four leaves, say a and b, such that $\overline{ap} = \alpha y$, $\overline{bp} = \beta y$, with $\alpha, \beta \leq \frac{1}{2}$. Let $\theta = $ angle apb. Let $x = $ dist (a, b). Then

$$x^2 = \alpha^2 y^2 + \beta^2 y^2 - 2\alpha\beta y^2 \cos \theta, \qquad (3)$$

$$x^2/y^2 = \alpha^2 + \beta^2 - 2\alpha\beta \cos \theta, \qquad (4)$$

from which it follows that

$$(\alpha - \beta)^2 < x^2/y^2 < (\alpha + \beta)^2. \qquad (5)$$

If at least one of the numbers $\alpha, \beta < \frac{1}{2}$, then $x^2/y^2 < 1$, which is impossible since $x \geq y$. If both α and $\beta = \frac{1}{2}$, then, since $1 - \cos \theta < 2$,

$$x^2/y^2 = \frac{1}{2}(1 - \cos \theta) < 1,$$

which is impossible. Therefore there must be a leaf at every intersection of a left parastichy and a right parastichy of the opposed parastichy pair determined by p and q.

That p and q are denominators of consecutive principal convergents of d is then a consequence of the following theorem:

Theorem 2. If either (p, q) or (q, p) is a visible opposed parastichy pair of a leaf distribution with divergence d, and both p and q are the denominators of principal convergents of the simple continued fraction expansion for d, then they are the denominators of consecutive principal convergents.

The proof is given in Appendix D.

7. How Three Leaves Nearest Zero Equidistant from Zero are Related

Suppose that p and q are leaves nearest zero on opposite sides of l_0 and that the point (d, r), as required by assumption (3), is confined to the (p, q) semicircle. As r decreases, the point (d, r) descends on the semicircle until r reaches a value at which another leaf s, with leaf number $s > p, q$, becomes a leaf nearest zero if s is already present. Then for every leaf $w \neq 0$

dist $(0, w) \geq$ dist $(0, p) =$ dist $(0, q) =$ dist $(0, s)$. We can now determine how s is related to the divergence d and to p and q. The fact that s is a leaf nearest zero implies that s is a point of close return and hence is the denominator of a principal convergent of the simple continued fraction for d. By section 6, p and q are the denominators of consecutive principal convergents. Let the smaller of them be q_{n-1}. Then the larger one is q_n. Since q_{n-1} and q_n are on opposite sides of l_0, s is on the same side with one of them and opposite the other. If s and q_{n-1} were on opposite sides of l_0, then, by section 6, we would have $s = q_n$, contradicting the assumption that s is greater than both q_{n-1} and q_n. Therefore s and q_n are on opposite sides of l_0. Then, by section 6, s must be q_{n+1}. That is, p, q and s are consecutive points of close return, or the denominators of *consecutive* principal convergents of d. Then, by equation (D1) of Appendix D,

$$q_{n+1} = q_{n-1} + a_{n+1} q_n,$$

where a_{n+1}, the $(n+1)$th term of the simple continued fraction for d, is a positive integer.

While (d, r) is on the (p, q) semicircle, dist $(0, p) =$ dist $(0, q)$. Since we have identified the smaller of the two numbers p and q as q_{n-1} and the larger as q_n, we have dist $(0, q_{n-1}) =$ dist $(0, q_n)$. At the moment when s first becomes capable of being a leaf nearest zero if it were already present, $s = q_{n+1}$ and dist $(0, q_{n-1}) =$ dist $(0, q_n) =$ dist $(0, q_{n+1})$. Then, in the plane development of the cylindrical representation of the leaf distribution, the three lattice points q_{n-1}, q_n and q_{n+1} lie on a circle whose center is 0, and q_{n-1} and q_{n+1} are on the same side of l_0.

If $a_{n+1} > 1$, the continued fraction for d has intermediate convergents $(p_{n-1} + a_j p_n)/(q_{n-1} + a_j q_n)$, $1 \leq a_j < a_{n+1}$, whose denominators are the leaf numbers of intermediate leaves lying on the chord that joins q_{n-1} and q_{n+1} (Coxeter, 1972). Then dist $(0, q_{n-1} + a_j q_n) <$ dist $(0, q_{n+1})$, which

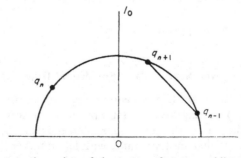

FIG.5. Three consecutive points of close return that are equidistant from 0.

contradicts the fact that for every $w \neq 0$ dist $(0, w) \geq$ dist $(0, q_{n+1})$. Therefore it is necessary that $a_{n+1} = 1$. Consequently,

$$s = q_{n+1} = q_{n-1} + q_n. \tag{6}$$

That is, the next point of close return after q_{n-1} and q_n is calculated by the recurrence relation that characterizes Fibonacci numbers.

We saw in Appendix D, in the proof of proposition D, that the confinement of the point (d, r) to the (p, q) semicircle determines finitely many integers $a_1, \ldots, a_n$ as terms of the simple continued fraction for d.

$$d = 0 + 1/a_1 + 1/a_2 + \ldots + 1/(a_n + x), \quad \text{with } x < 1.$$

Hence $d = 0 + 1/a_1 + 1/a_2 + \ldots + 1/a_n + 1/(a_{n+1} + y)$, where $a_{n+1} = [1/x]$, $y < 1$. We have just shown that when (d, r) reaches the position where a leaf $s > p$, q first can become leaf nearest zero (if it is present) with dist $(0, s) =$ dist $(0, p) =$ dist $(0, q)$, then the value of one more term of the continued fraction is determined, namely, $a_{n+1} = 1$.

The (p, q) semicircle, as we have seen, is the (q_{n-1}, q_n) semicircle. Let Q be the intersection of the (q_{n-1}, q_n) semicircle and the (q_n, q_{n+1}) semicircle in the (d, r) plane, where $q_{n+1} = q_{n-1} + q_n$, and let T_Q be the time when the point (d, r) reaches Q as (d, r) slides down the (q_{n-1}, q_n) semicircle. There are two cases that may arise: Case I: $T_Q \geq q_{n+1}$. Case II: $T_Q < q_{n+1}$. In case I, leaf q_{n+1} is present when the point (d, r) reaches Q, and, when (d, r) moves below Q, q_{n+1} replaces q_{n-1} as leaf nearest zero. (q_{n+1} replaces q_{n-1} rather than q_n as leaf nearest zero because q_{n+1} and q_{n-1} are on the same side of l_0.) So, as r continues to fall, the point (d, r) switches from the (q_{n-1}, q_n) semicircle to the (q_n, q_{n+1}) semicircle. In case II, the leaf q_{n+1} emerges after the point (d, r) has passed Q. Under our assumption that the minimum distance between leaves continues to be maximized, the point (d, r) leaps horizontally from the (q_{n-1}, q_n) semicircle to the (q_n, q_{n+1}) semicircle. [This discontinuity in the path of (d, r) in the model results from assuming that the leaf q_{n+1} is fully grown at birth. In a plant, where a new leaf attains its full growth gradually, this discontinuity would not exist: (d, r) would follow a transitional path from one semicircle to the other.]

Definition. Suppose, when contact pressure begins, the state of the system is on the (q_{n-1}, q_n) semicircle. The *contact pressure path* that starts on the (q_{n-1}, q_n) semicircle is a sequence of arcs joined to each other in succession: first the arc of the (q_{n-1}, q_n) semicircle down to its intersection with the (q_n, q_{n+1}) semicircle, where $q_{n+1} = q_{n-1} + q_n$; then the arc of the (q_n, q_{n+1}) semicircle from that point to its intersection with the (q_{n+1}, q_{n+2}) semicircle, where $q_{n+2} = q_n + q_{n+1}$; and so on.

Examples. The contact pressure path that starts with the $(1, 2)$ semicircle is made up of arcs of the $(1, 2)$, $(2, 3)$, $(3, 5)$, $(5, 8)$ semicircles, etc., joined in succession. This path was called the *normal phyllotaxis path* in Adler (1974). The contact pressure path that starts with the $(2, 5)$ semicircle is made up of arcs of the $(2, 5)$, $(5, 7)$, $(7, 12)$ semicircles, etc., joined in succession. This path was called the *anomalous phyllotaxis path* in Adler (1974).

Repeated application of the argument of this section establishes the following theorem.

Theorem 3. For a leaf distribution with $1/(t+1) < d < 1/t$, once (d, r) is on the contact pressure path that starts on the (q_{n-1}, q_n) semicircle, then, with continuing contact pressure, it either switches from one arc of the path to the next at their intersection, or, if it overshoots the intersection, jumps back horizontally to a position near or on another of the arcs when the appropriate leaf emerges.

Since (q_{n-1}, q_n) or (q_n, q_{n-1}) is attainable by a sequence of extensions of $(0, 1)$, namely

$a_1 = t$ left extensions followed by

a_2 right extensions followed by

a_3 left extensions followed by

$$\cdot \quad \cdot \quad \cdot$$

$$a_n \begin{pmatrix} \text{left if } n \text{ is odd} \\ \text{right if } n \text{ is even} \end{pmatrix} \text{extensions}$$

that end with (q_{n-1}, q_n) or (q_n, q_{n-1}), then

$$\lim_{T \to \infty} d = 1/t + 1/a_1 + \ldots + 1/a_n + 1/\tau.$$

For the case $t = 2$,

$$\lim_{T \to \infty} d = 1/2 + 1/a_2 + \ldots + 1/n_n + 1/\tau.$$

8. The Consequences of Contact Pressure

We can now say precisely what the consequences of contact pressure are. When contact pressure begins at $T = T_c$, the value of T_c and the initial values of d and r determine two leaf numbers p and q such that maximization of the minimum distance between leaves requires that (d, r) be on the (p, q)

semicircle where dist $(0, p) = $ dist $(0, q)$, and p and q are on opposite sides of l_0. Then, as r decreases to zero, the point (d, r) descends on the contact pressure path that starts on the (p, q) semicircle, though it may leave it temporarily for short periods of time. If p is the smaller of the two leaf numbers p and q, the leaf distribution advances to higher and higher phyllotaxis only via the phyllotaxis expressed by successive pairs of consecutive terms of the Fibonacci sequence generated by the numbers p and q, namely,

$$p, q, p+q, p+2q, 2p+3q, \ldots . \tag{7}$$

The divergence, starting from some value

$$d = 1/t + 1/a_2 + \ldots + 1/(a_n + x), \quad \text{with } x < 1,$$

then converges to an "ideal" angle

$$1/t + 1/a_2 + \ldots + 1/a_n + 1/\tau. \tag{8}$$

The zig-zag shape of a contact pressure path suggests the use of the following metaphor to describe what happens under contact pressure. There are many vortices in the (d, r) plane each of which is a contact pressure path. Each vortex begins with an arc of a semicircle where min dist^2 $(0, n)$ is maximized. The semicircle itself is determined by a maximum in the graph of min dist^2 $(0, n)$ as a function of d for given T and r. Figure 3 shows that for low T and high r there is only one such maximum. But the number of maxima increases as T increases and r decreases. Hence with increasing T and decreasing r the number of vortices in the plane increases. When contact pressure begins, wherever the point (d, r) may be in the (d, r) plane at that moment, it is immediately drawn to the nearest attainable vortex. (Attainable means able to be reached by a change in d that increases the minimum distance between leaves.) Then, as r decreases, (d, r) descends into the vortex, perhaps sometimes leaving it temporarily for a short time, but always being drawn back.

We have as an immediate corollary of this general result a necessary and sufficient condition for contact pressure to produce normal phyllotaxis.

Theorem 4. A necessary and sufficient condition for contact pressure to produce normal phyllotaxis is that, when contact pressure begins, the leaf numbers p and q of the (p, q) semicircle to which the point (d, r) is then confined be consecutive Fibonacci numbers of the sequence

$$1, 2, 3, 5, 8, \ldots . \tag{9}$$

I. ADLER

9. The First Few (p, q) Semicircles

In order to determine an important *sufficient* condition for contact pressure to produce normal phyllotaxis, we examine the first few (p, q) semicircles that are possible when $T \le 5$, in the region of the (d, r) plane where $1/3 < d < 1/2$. We subdivide the interval $(1/3, 1/2)$ by putting in some successive mediants. First put in 2/5, the mediant between 1/3 and 1/2. Then put in 3/8, the mediant between 1/3 and 2/5; and put in 3/7, the mediant between 2/5 and 1/2. Table 1 shows the ranges of d that are obtained in this way, the simple continued fraction for d in each range, and the principal convergents and points of close return that are determined in each range. Note that the smaller the range, the more points of close return are determined.

TABLE 1

Range of d	Continued fraction	Principal convergents	Points of close return
$1/3 < d < 1/2$	$d = 1/2 + 1/1 + x$ $(0 < x < \infty)$	0/1, 1/2	1, 2
$1/3 < d < 2/5$	$d = 1/2 + 1/1 + 1/1 + x$ $(0 < x < \infty)$	0/1, 1/2, 1/3	1, 2, 3
$2/5 < d < 1/2$	$d = 1/2 + 1/2 + x$ $(0 < x < \infty)$	0/1, 1/2	1, 2
$2/5 < d < 3/7$	$d = 1/2 + 1/2 + 1/1 + x$ $(0 < x < \infty)$	0/1, 1/2, 2/5	1, 2, 5
$3/8 < d < 2/5$	$d = 1/2 + 1/1 + 1/1 + 1/1 + x$ $(0 < x < \infty)$	0/1, 1/2, 1/3, 2/5	1, 2, 3, 5

Note. In the interval $1/3 < d < 2/5$, dist $(l_0, 4) > 1/3$, while dist $(l_0, 2) < 1/3$. In the interval $2/5 \le d < 1/2$, dist $(l_0, 4) = 2$ dist $(l_0, 2)$. Therefore, for $T \le 5$, and $1/3 < d < 1/2$, leaf 4 is not a point of close return.

The possible (p, q) semicircles up to $T = 5$ are (1, 2), (2, 3), (2, 5) and (3, 5). (1, 3) is excluded because 1 and 3 are not consecutive points of close return, and indeed, 1 and 3 are on the same side of l_0. For $T < 3$, only the (1, 2) semicircle is relevant, since leaves 3 and 5 do not exist then. For $T < 5$, the (1, 2) and (2, 3) semicircles are relevant (see Fig. 6). For $T < 6$, the (1, 2), (2, 3), (2, 5) and (3, 5) semicircles are relevant. However, leaf 5 becomes a leaf nearest zero only when (d, r) is below the common point of intersection of the (2, 3), (2, 5) and (3, 5) semicircles. The value of r at this

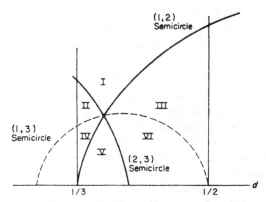

FIG. 6. Distances from leaf 0 if (d, r) is in the region indicated:

    I. dist $(0, 1) <$ dist $(0, 2) <$ dist $(0, 3)$.
    II. dist $(0, 1) <$ dist $(0, 3) <$ dist $(0, 2)$.
    III. dist $(0, 2) <$ dist $(0, 1) <$ dist $(0, 3)$.
    IV. dist $(0, 3) <$ dist $(0, 1) <$ dist $(0, 2)$.
    V. dist $(0, 3) <$ dist $(0, 2) <$ dist $(0, 1)$.
    VI. dist $(0, 2) <$ dist $(0, 3) <$ dist $(0, 1)$.

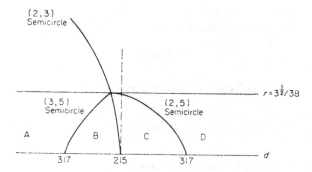

FIG. 7. Distances from leaf 0 if (d, r) is in the region indicated:

    A. dist $(0, 3) <$ dist $(0, 5) <$ dist $(0, 2)$.
    B. dist $(0, 5) <$ dist $(0, 3) <$ dist $(0, 2)$.
    C. dist $(0, 5) <$ dist $(0, 2) <$ dist $(0, 3)$.
    D. dist $(0, 2) <$ dist $(0, 5) <$ dist $(0, 3)$.

point of intersection is obtained by solving the pair of simultaneous equations

$$(d-1/5)^2+r^2 = (1/5)^2, \quad r > 0;$$
$$(d-7/16)^2+r^2 = (1/16)^2, \quad r > 0; \tag{10}$$

which represent the (2, 3) and (3, 5) semicircles respectively. We get for this intersection

$$r = 3^{\frac{1}{2}}/38. \tag{11}$$

Therefore the (2, 5) and (3, 5) semicircles are relevant only for $r \leq 3^{\frac{1}{2}}/38$ (see Fig. 7). If a horizontal line $r = c$ is drawn in Fig. 7, its intersection with each semicircle indicates a value of d at which min dist^2 $(0, n)$ is maximized for $r = c$ and T large enough for the semicircle to be relevant. Note the relation of condition (10) to the number of maxima: If $r \geq 3^{\frac{1}{2}}/38$, and T is arbitrary but ≥ 2, there is one maximum; if r is arbitrary, and $T < 5$, there is one maximum; if $r < 3^{\frac{1}{2}}/38$ and $T = 5$, there are two maxima; if $r < 3^{\frac{1}{2}}/38$ and $T > 5$, there are two or more maxima.

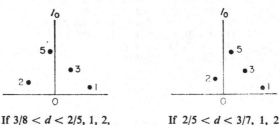

If $3/8 < d < 2/5$, 1, 2, 3 and 5 are points of close return.

If $2/5 < d < 3/7$, 1, 2 and 5 (but not 3) are points of close return.

FIG. 8.

10. A Sufficient Condition for Normal Phyllotaxis

The minimum of dist $(0, n)$, $n \leq [T]$, plotted as a function of d for a given value of r, has only one maximum under either of the following conditions:

(1) $T < 5$, no matter what the value of r;
(2) $r \geq 3^{\frac{1}{2}}/38$, no matter what the value of T.

If contact pressure begins under either of these conditions, then, in either case, the point (d, r) at which min dist $(0, n)$ is maximized is on either the (1, 2) semicircle or the (2, 3) semicircle, and both pairs (1, 2) and (2, 3) are consecutive terms of the Fibonacci sequence (9). Consequently we have the following.

Theorem 5. A sufficient condition for normal Fibonacci phyllotaxis is that either $T_c < 5$ or $r(T_c) \geq 3^{\frac{1}{2}}/38$. That is, *normal Fibonacci phyllotaxis is inevitable if contact pressure begins early,* where early means either before leaf 5 emerges or before r has fallen below $3^{\frac{1}{2}}/38$.

11. Computer Verification of the Theorems

For any given function $r(T)$ that decreases monotonically to 0, the values of d that correspond to each value of r under conditions of contact pressure can be calculated by computer, and the corresponding phyllotaxis can be determined. The tables in Appendix E show the calculated results for selected functions $r(T)$ and selected initial conditions (values of T_c and initial d). The calculations were made on a Hewlett–Packard HP-65 programmable calculator.

Tables 4 and 5 use $r = 1 \cdot 5 T^{-2}$. In Table 4, contact pressure begins at $T = 3$, with d initially between 1/3 and 1/2. Here d is converging to $\tau^{-2} = 0 \cdot 381966012$ (normal phyllotaxis). In Table 5, contact pressure begins at $T = 6$, with d initially between 2/5 and 1/2. Here d is converging to $(7 + 5^{\frac{1}{2}})/22 = 0 \cdot 419821272$ (anomalous phyllotaxis).

In Table 6, $r = T^{-1}/4$. In Table 7, $r = T^{-0 \cdot 3}/15$. In both cases, contact pressure begins when $T = 2$, with d initially between 1/3 and 1/2. The result is normal phyllotaxis.

In Tables 8 and 9,

$$r = (2\pi)^{-1} \ln \{(1 + aT)/[1 + a(T-1)]\},$$

with $a = 0 \cdot 14$, and contact pressure starting early. In Table 8, dist $(0, n)$ was calculated assuming $r_i = r$ for $i = 1, \ldots, n$. In Table 9, it was assumed that the r_i were all different, and in fact that $r_i(T) = r_1(T - i + 1) = r(T - i + 1)$. Although the values of d in the two tables are initially slightly different, they converge toward each other with increasing T, and both yield normal phyllotaxis. The formula for r used in the tables was derived from the model of the sunflower described in Appendix A.

In Tables 10 and 11,

$$r = (4\pi)^{-1}[1 + 4a(T + b)]^{\frac{1}{2}} \ln [(T + b)/(T + b - 1)],$$

with $a = 0 \cdot 1$ and $b = 7$. This value of r was derived from the parabolic model of Appendix B. In Table 10, contact pressure starts at $T = 2$, with d initially between 1/3 and 1/2, and the phyllotaxis is normal. In Table 11, contact pressure starts at $T = 5$, with d initially between 2/5 and 1/2, and the phyllotaxis is anomalous.

12. Equality of Successive Divergences is a Consequence of Contact Pressure

In the model constructed here we have assumed that $d_i(T) = d(T)$ for all $i \leq [T]$. Actually this assumption can be replaced by the weaker assumption that successive primordia merely lie on a single fundamental spiral. Then the d determined by the equation for a (p, q) semicircle,

$$p^2 r^2 + [pd - X(pd)]^2 = q^2 r^2 + [qd - X(qd)]^2, \tag{12}$$

is such that it is simultaneously the average of $d_1, \ldots, d_p$ and the average of $d_1, \ldots, d_q$. Moreover, since each leaf in turn plays the role of leaf 0, d is simultaneously the average of $d_n, \ldots, d_{n+p-1}$ and the average of $d_n, \ldots, d_{n+q-1}$ for any n. With this condition as hypothesis we can prove that contact pressure compels an equalization of the d_i. The proof is in Appendix F.

A consequence of this is that a complete model of phyllotaxis is obtained when the contact pressure model developed here is combined with the weakest form of a space-filling model which suffices to assure that successive primordia generate a fundamental spiral (see Adler, 1975).

13. Amendments to Adler (1974)

The present paper is both a generalization and a correction of Adler (1974) as follows.

(1) It was asserted in Adler (1974, pp. 20–21) without proof that when maximization of the minimum distance between leaves puts (d, r) on a (p, q) semicircle, that p and q are consecutive points of close return for the corresponding value of d. The proof has been supplied here in section 6 and Appendix D.

(2) It was implied in Adler (1974, pp. 20–21) that when the point (d, r) migrates from its initial position at $T = T_c$ to the (p, q) semicircle, it does so from a region where the leaf nearest zero is the greater of the two leaf numbers p, q. This is not correct. The initial leaf nearest zero before (d, r) migrates to the (p, q) semicircle may be either one of the leaf numbers p, q.

(3) It was asserted in Adler (1974, p. 7) that the formula $R = e^{2\pi r}$ suffices for transforming a disc model into its normalized cylindrical representation. However, it should be noted that the plastochrone ratio measured in a transverse section of a stem apex whose surface is a surface of revolution is not the plastochrone ratio of the corresponding disc model. The latter is the equivalent plastochrone ratio obtained from the former by multiplying by an appropriate correction factor, as explained in Appendix C. The

correct formula, then, for obtaining the rise r of a cylindrical representation of a zone of a surface of revolution, when R is the plastochrone ratio of a transverse section, is $r = (2\pi \sin \alpha)^{-1} \ln R$, where α is the angle the surface of the zone, assumed to be conical, makes with the axis of the surface of revolution.

(4) In Adler (1974) the condition $1\cdot14 \leq T^2 r(T) < 1\cdot79$ was assumed in the derivation of a sufficient condition for contact pressure to produce normal phyllotaxis. This condition has now been replaced by the more general assumption that r is a monotonic decreasing function of T with

$$\lim_{T \to \infty} r(T) = 0.$$

It is clear that the earlier condition has no biological significance whatever. Its significance was purely heuristic, in that it led to the discovery of the normal phyllotaxis path and the more general sufficient condition for it established in this paper.

(5) In Adler (1974) conspicuous opposed parastichy pairs were defined only in the case where the opposed parastichy pairs are visible. Under conditions of contact pressure, this restriction, though correct, is unnecessary. In section 6 of this paper it was proved that, if a conspicuous opposed parastichy pair is defined merely as one determined by two leaves nearest zero on opposite sides of l_0, then, where contact pressure exists, "conspicuous" implies "visible".

(6) In Adler (1974) it was asserted that Schwendener failed in his attempt (1878) to show that contact pressure can explain normal phyllotaxis "because he unconsciously assumed what he was trying to prove when he assumed without proof that the successive opposed parastichy pairs that are conspicuous are (F_n, F_{n-1}) with n increasing with time". This criticism is oversimplified. A more accurate appraisal of what Schwendener did is given in section 14.

(7) Equation (22), p. 21 in Adler (1974) has a typographical error. The corrected equation is

$$[d - (p_i q_i - p_{i-1} q_{i-1})/(q_i^2 - q_{i-1}^2)]^2 + r^2 = 1/(q_i^2 - q_{i-1}^2)^2, \quad r > 0.$$

14. Schwendener Re-examined

The model of contact pressure developed here and in Adler (1974) and the rigorous deduction of its consequences shows that Schwendener was right in his belief that contact pressure can account for the advance to higher phyllotaxis by the replacement of (p, q) phyllotaxis, $p < q$, by $(q, p+q)$

phyllotaxis, and for the convergence of the divergence to the ideal angle τ^{-2} when the initial (p, q) is $(1, 2)$. Although his proof (Schwendener, 1878) is faulty, he did obtain an important part of the result of theorem 8 in Adler (1974). Schwendener's failures and accomplishments in his part I, chapter 1, where his principal results were presented, are outlined below.

(1) Schwendener assumes, when p and q are leaves nearest zero on opposite sides of l_0, that dist $(0, p) =$ dist $(0, q)$. He gives no justification for this assumption other than that it is a simplifying assumption. However, we have shown that it is more than that. We have shown that it is a consequence of the presence of contact pressure represented here by assumption (3).

(2) Schwendener pictures the opposed parastichy triangle determined by the two leaves nearest zero as a gabled roof formed by sloping rafters in the plane development of the cylindrical surface. He then "proves" that a vertical force of compression applied at the vertex of the gable causes the feet of the rafters to move apart. His "proof" consists of first resolving the vertical force into two oblique components, and then resolving each of the latter into vertical and horizontal components. He arrives by this circuitous path to a resolution of his original vertical force into a vertical force plus two horizontal forces of equal magnitude but with opposite directions, one pushing the foot of the left rafter to the left, and the other pushing the foot of the right rafter to the right. But, although these two points are seen as distinct points in the plane development of the cylindrical surface, they really represent the same point on the cylinder. The two forces of equal magnitude and opposite direction applied at the same point therefore add up to zero (as indeed they should) and cannot serve to explain any motion whatever.

(3) Nevertheless, a separation of the two feet of the rafters does indeed take place if the girth of the cylinder is increasing. Since Schwendener does assume such an increase, he is justified in concluding that the separation takes place, but the separation has nothing whatever to do with the assumed vertical force of compression.

(4) The assumption that the girth of the cylinder is increasing while the distance from leaf 0 to the leaves nearest zero remains the same and their leaf numbers remain the same (until one of them is displaced by another leaf nearest zero) implies that the rise r (normalized internode distance) is decreasing. This decreasing rise, combined with the equality of the distances from leaf zero of the leaves nearest zero means that the conditions Schwendener assumes satisfy the hypotheses on the basis of which theorem 3 in section 7 of this paper was proved. Thus Schwendener's conclusion that under these conditions (p, q) phyllotaxis with $p < q$ gives way to $(q, p+q)$ phyllotaxis is correct.

(5) Schwendener proceeds to determine the consequences of these conditions in two different ways: (i) by a sequence of ruler and compass constructions; (ii) by using an apparatus employing rollers confined in a frame made of a hinged parallelogram (like a draftsman's parallel rulers) and with constraints imposed on some of the rollers by grooves into which the axles of these rollers are inserted. He then observes, both from his constructions and from the operation of the apparatus, that (1, 2) phyllotaxis gives way to (2, 3) phyllotaxis, which in turn gives way to (3, 5) phyllotaxis, etc., the sequence being precisely the sequence of pairs obtained by taking successive consecutive terms of the Fibonacci sequence 1, 2, 3, 5, 8, ' . . . These observations have the disadvantage that they are not deduced by logical argument from his premises but are only empirical observations. However, they have the virtue of being empirical observations made with the help of an apparatus that is a valid analog computer that accurately conforms to the assumed conditions.

(6) He does prove on the basis of these empirical observations that the divergence alternately decreases and increases, with narrower and narrower swings, and converges to the ideal angle.

(7) He also calculates the angles at which the oscillating divergence changes direction and the transition is made to higher phyllotaxis. Although his method of calculation is based on using approximations and empirical observations, he comes remarkably close to the values of these angles predicted by the formula for d_n in theorem 8 of Adler (1974). The predictions of theorem 8, calculated to the nearest second of arc, and Schwendener's results, given in degrees and minutes, are shown in Table 2.

(8) Schwendener's derivation does not reveal that contact pressure makes d a function of r.

TABLE 2

Limiting angles

n	d_n from theorem 8, in Adler (1974)	As given by Schwendener (1878)
3	180°	180°
4	128° 34′ 17″	128° 34′
5	142° 6′ 19″	142° 6′
6	135° 55′ 6″	135° 55′
7	138° 8′ 22″	138° 8′
8	137° 16′ 12″	137° 16′
9	137° 35′ 57″	137° 36′
10	137° 28′ 22″	137° 26′

I. ADLER

(9) It is this functional relationship between d and r, predicted in theorem 8, that now opens the way to an observational test of the contact pressure theory.

I am grateful to Peggy Adler for the figures.

REFERENCES

ADLER, I. (1974). *J. theor. Biol.* **45**, 1.
ADLER, I. (1975). *J. theor. Biol.* **53**, 435.
COXETER, H. S. M. (1961). *Introduction to Geometry*, p. 327. New York: Wiley.
COXETER, H. S. M. (1972). *J. Algebra* **20**, 167.
DAVIES, P. A. (1939). *Am. J. Bot.* **26**, 67.
MATHAI, A. M. and DAVIS, T. A. (1974). *Math. Biosci.* **20**, 117.
RICHARDS, F. J. (1948). *Symp. Soc. exp. Biol.* **2**, 217.
RICHARDS, F. J. (1951). *Phil. Trans. R. Soc. B* **235**, 509.
SCHOUTE, J. C. (1913). *Recl. Trav. bot. néerl.* **10**, 153.
SCHWENDENER, S. (1878). *Mechanische Theorie der Blattstellungen*. Leipzig: Engelmann.

APPENDIX A

Cylindrical Representation of a Leaf Distribution on a Disc

Representation of a leaf distribution as points on a disc arises naturally in two ways. (i) Some distributions, e.g. the florets on the capitulum of *helianthus, are* on a surface that is approximately a disc. (ii) The leaf centers in a transverse section of the growing tip of a stem are on a disc. The section itself is not a disc representation, but can be transformed into one as explained in Appendix B.

In a cylindrical model normalized by taking the girth of the cylinder as unit of length, let (x_i, y_i) be the co-ordinates of leaf i relative to horizontal and vertical axes through leaf 0 in the plane development of the cylindrical surface. Then $0 \leq x_i < 1$. Let θ_i be the angle subtended by x_i at the axis of the cylinder. Then

$$\theta_i = 2\pi x_i; \qquad d_i = x_i - x_{i-1} \quad \text{if } x_i > x_{i-1}$$

and

$$d_i = 1 + x_i - x_{i-1} \quad \text{if } x_i < x_{i-1}; \qquad r_i = y_i - y_{i-1}.$$

In a disc model, leaf 0 is on the circumference of the disc. Polar co-ordinates (ρ_i, θ_i) of leaf i can be introduced by using the center of the disc as origin and the radius to leaf 0 as axis.

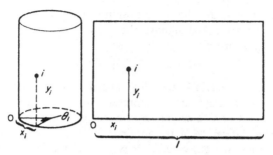

FIG. 9. Normalized cylindrical model (girth = 1).

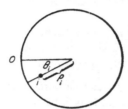

FIG. 10. Disc model.

A cylindrical representation of the disc model that preserves ortho-gonality of parastichies is given by

$$\rho_i = (2\pi)^{-1} e^{-2\pi y_i}; \qquad \theta_i = 2\pi x_i.$$

If R_i is the plastochrone ratio for leaves $i-1$ and i in the disc model, and r_i is the corresponding rise in the cylindrical model, then

$$R_i = \rho_{i-1}/\rho_i = e^{-2\pi(y_{i-1}-y_i)} = e^{2\pi r_i}.$$

Consequently,

$$r_i = (2\pi)^{-1} \ln R_i. \qquad (A1)$$

The divergence d_i between leaves $i-1$ and i is the same in both models.

If R_i and r_i vary with time T, we may write $R_i = R_i(T)$ and $r_i = r_i(T)$ to indicate the corresponding functions of T. Then equation (A1) becomes

$$r_i(T) = (2\pi)^{-1} \ln R_i(T). \qquad (A2)$$

In the disc, if a circle with its center at the center of the disc is drawn through each leaf i, the disc is divided into a series of concentric rings each of which separates two consecutive leaves. On the cylinder, if a circle is drawn through each leaf in the plane that is perpendicular to the axis of the cylinder, the cylindrical surface is divided into a series of zones. Each ring in the disc, with an associated plastochrone ratio $R_i(T)$, corresponds to a zone on the cylinder with width $r_i(T)$. If successive plastochrone ratios

$R_i(T)$ are equal, then so are the corresponding rises $r_i(T)$. Then we may drop the subscripts and simply designate them as $R(T)$ and $r(T)$ respectively, related by the equation

$$r(T) = (2\pi)^{-1} \ln R(T). \tag{A3}$$

This equation applies trivially to a leaf distribution consisting of only two leaves. Consequently, a cylindrical representation of a leaf distribution in a disc may be thought of as a sequence of zones one on top of the other, each obtained from equation (A3) to obtain a cylindrical representation of a single ring between two consecutive leaves. This method of piecewise construction, zone by zone, of a cylindrical representation of a leaf distribution will be used in the example below and in Appendix B.

Example: a model of a sunflower. Richards (1948) and Mathai & Davis (1974) used a disc model of a sunflower in which radii to successive florets are assumed to be in arithmetic progression.

Let c = the length of the radius to a floret at the moment of its emergence;
    b = the increment of that length per plastochrone of time as the floret moves away from the center;
$A(i)$ = the age of leaf i in plastochrones;
    T = the time measured in plastochrones, with $T = 0$ when $A(0) = 0$.

For each leaf $i \geq 0$ at time T,

$$i + A(i) = T. \tag{A4}$$

$$A(i+1) = A(i) - 1. \tag{A5}$$

If $\rho_i(T)$ is the radius to leaf i at time T,

$$\rho_i(T) = c + bA(i). \tag{A6}$$

Then

$$R_{i+1}(T) = \rho_i(T)/\rho_{i+1}(T) = [c + bA(i)]/[c + bA(i+1)]$$
$$= [1 + aA(i)]/\{1 + a[A(i) - 1]\}$$

where $a = b/c$. Then

$$R_{i+1}(T) = [1 + a(T-i)]/[1 + a(T-i-1)]. \tag{A7}$$

For $i = 0$ we have

$$R_1(T) = (1 + aT)/[1 + a(T-1)]. \tag{A8}$$

Also

$$R_{i+1}(T) = R_1(T-i). \tag{A9}$$

Thus, the successive values of $R_i(T)$ obtained as i increases from 1 correspond to the successive values of $R_1(T)$ as T decreases to 0, so that the variation in $R_i(T)$ from the center of the disc outward recapitulates the history of $R_1(T)$ at the rim.

To obtain a cylindrical representation of this disc model of a sunflower, we build it up zone by zone from the cylindrical representation of the concentric rings that separate consecutive florets in the disc. As we have seen, divergences d_i are carried over from the disc to the cylinder without change. We shall designate these as $d_i(T)$ to indicate any functional dependence on time. The rises $r_{i+1}(T)$ that correspond to successive plastochrone ratios $R_{i+1}(T)$ are obtained from (A2) and (A7):

$$r_{i+1}(T) = (2\pi)^{-1} \ln \{[1+a(T-i)]/[1+a(T-i-1)]\}. \qquad (A10)$$

To simplify the notation slightly let us write simply $r(T)$ for $r_1(T)$. Then we have

$$r(T) = (2\pi)^{-1} \ln \{[1+aT]/[1+a(T-1)]\}. \qquad (A11)$$

Note that the first derivative of $r(T)$ with respect to T is negative, so that this model of the sunflower satisfies assumption (2) of section 1.

APPENDIX B

Cylindrical Representation of a Leaf Distribution on a Surface of Revolution

The surface of the growing tip of a stem is, typically, a surface of revolution, usually described as being approximately parabolic. To obtain a cylindrical representation of a leaf distribution on such a surface, the surface is first divided into zones by drawing a circle through each leaf in the plane that is perpendicular to the axis of the surface. Then each of the zones is transformed into a zone on a normalized cylindrical surface. The transformation is carried out in two steps: (i) each zone is transformed into a ring in a disc representation; (ii) the representation in a disc is transformed into a cylindrical representation via equation (A2).

Step (i). The zone between leaves i and $i+1$ is closely approximated by a conical surface joining the circles that are its edges, so it is replaced by this conical surface. (This is analogous to the use of the Lambert conformal conic projection used to make maps for air navigation.) Project leaf $i+1$ and the circle through it onto the plane of the circle through leaf i. Let $(i+1)'$ be the projection of $i+1$. The same projection, extended to all points of the zone of the conical surface, transforms it into a ring in a disc, as shown in Fig. 11. Phyllotaxis relations in this disc are essentially those found

56   I. ADLER

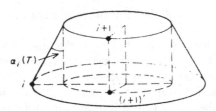

FIG. 11. Projection of a zone of a conical surface into a transverse disc.

in a transverse section of the stem taken through leaf i. Let $d_{i+1}(T)$ and $R_{i+1}(T)$ be the divergence and the plastochrone ratio between leaves i and $(i+1)'$ in the transverse section through i at time T. Let $\alpha_i(T)$ be the angle at time T between the axis of the cone and the element of the conical surface. For a narrow zone this element almost coincides with a line tangent to the surface of revolution at leaf i in the plane determined by i and the axis of the surface of revolution. So, for computational purposes, we take $\alpha_i(T)$ to be the angle between this tangent line and the axis.

As shown by Richards (1951), the disc picture obtained by a transverse section does not give a true representation of parastichy relations on the surface of the stem. To obtain such a true representation with $d_{i+1}(T)$ as divergence, it is necessary to replace the plastochrone ratio $R'_{i+1}(T)$ by the equivalent plastochrone ratio $R_{i+1}(T)$, where $R'_{i+1}(T)$ and $R_{i+1}(T)$ are related by the equation

$$\ln R_{i+1}(T) = [\sin \alpha_i(T)]^{-1} \ln R'_{i+1}(T). \tag{B1}$$

A derivation of this equation based on a simple intuitive geometric argument is given in Appendix C.

Step (ii). The cylindrical representation of the zone is obtained by combining equations (A2) and (B1). Then we have

$$r_{i+1}(T) = [2\pi \sin \alpha_i(T)]^{-1} \ln R'_{i+1}(T). \tag{B2}$$

Note 1. When $\alpha_i(T) = \pi/2$ for all T, the surface of revolution becomes a disc and equation (B2) reduces to equation (A2).

Note 2. Coxeter (1961) gives a family of transformations depending on an arbitrary parameter b for transforming to a cylinder with radius a a cone whose radius is a at a distance c from the vertex. For a normalized cylinder the Richards transformation is a special case of the Coxeter transformation with $b = [(2\pi)^{-2} + c^2]^{\frac{1}{2}}$.

Example. A model of a parabolic stem dome with uniform vertical growth rate. (Note: this model is constructed only to have a specific parabolic model for which the consequences of contact pressure can be determined by computation. It is not claimed that this model represents any particular plant stem.)

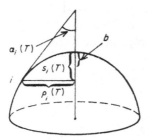

Fig. 12. A parabolic dome.

Assumptions. (1) Each new leaf emerges at a distance b below the top of the dome. (2) The dome grows one unit in height per plastochrone. (3) Let a line tangent to the dome at leaf i in the plane determined by i and the axis of the paraboloid make an angle $\alpha_i(T)$ at time T with the axis. Let $\rho_i(T)$ be the radius to leaf i at time T in a transverse section through i. Let $s_i(T)$ be the vertical distance of leaf i from the top of the dome at time T. We assume that $s_i(T)$ and $\rho_i(T)$ are related by the equation for a parabola:

$$s_i(T) = a\rho_i^2(T), \quad a > 0. \tag{B3}$$

From assumptions (1) and (2),

$$s_i(T) = T + b - i. \tag{B4}$$

Then

$$T + b - i = a\rho_i^2(T) \tag{B5}$$

and, in particular,

$$T + b = a\rho_0^2(T). \tag{B6}$$

Then

$$\rho_0(T-i) = \rho_i(T), \tag{B7}$$

$$R'_{i+1}(T) = [\rho_i(T)]/[\rho_{i+1}(T)] = [\rho_0(T-i)]/[\rho_0(T-i-1)], \quad i \geq 0, \tag{B8}$$

$$R'_1(T) = [\rho_0(T)]/[\rho_0(T-1)], \tag{B9}$$

$$R'_1(T-i) = [\rho_0(T-i)]/[\rho_0(T-i-1)] = R'_{i+1}(T). \tag{B10}$$

Therefore, as in the sunflower model described in Appendix A, the variation of $R'_{i+1}(T)$ from the youngest leaf to the oldest leaf recapitulates the history of $R_1'(T)$. From equations (B8) and (B5),

$$R_{i+1}'(T) = ([T+b-i]/[T+b-i-1])^{\frac{1}{2}}. \tag{B11}$$

Combining equation (B11) with equation (B1) we get

$$\ln R_{i+1}(T) = [\sin \alpha_i(T)]^{-1} \ln ([T+b-i]/[T+b-i-1])^{\frac{1}{2}}. \tag{B12}$$

Combining equation (B12) with equation (A2) we get

$$r_{i+1}(T) = [4\pi \sin \alpha_i(T)]^{-1} \ln ([T+b-i]/[T+b-i-1]). \tag{B13}$$

To evaluate $[\sin \alpha_i(T)]^{-1}$, we note that in Fig. 13, which is the same as Fig. 12 upside down, the slope of the line tangent to the paraboloid at leaf i

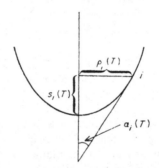

FIG. 13. A vertical section of a parabolic dome, drawn upside-down.

in the plane determined by i and the axis of the paraboloid is given by the derivative of $s_i(T)$ with respect to $\rho_i(T)$, but it is also equal to $\cot \alpha_i(T)$. Then

$$\cot \alpha_i(T) = 2a\rho_i(T) = 2[a(T+b-i)]^{\frac{1}{2}}.$$

Then, since

$$[\sin \alpha_i(T)]^{-1} = [1+\cot^2 \alpha_i(T)]^{\frac{1}{2}},$$

we have

$$[\sin \alpha_i(T)]^{-1} = [1+4a(T+b-i)]^{\frac{1}{2}}. \tag{B14}$$

Combining equations (B14) and (B13) yields

$$r_{i+1}(T) = (4\pi)^{-1}[1+4a(T+b-i)]^{\frac{1}{2}} \ln ([T+b-i]/[T+b-i-1]). \tag{B15}$$

To simplify the notation we write $r(T)$ for $r_1(T)$. Then we have

$$r(T) = (4\pi)^{-1}[1+4a(T+b)]^{\frac{1}{2}} \ln ([T+b]/[T+b-1]). \tag{B16}$$

Note that in this model, as in the sunflower model of Appendix A, the derivative of $r(T)$ with respect to T is negative, so that this model of a leaf distribution on a parabolic dome satisfies assumption (2) of section 1.

APPENDIX C

Derivation of the Formula for Equivalent Plastochrone Ratio

Parastichy relations on a conical surface are truly represented by those in a plane development of the surface. Let ρ_i and ρ_{i+1} be the radii to leaves i and $i+1$ respectively in the transverse section of the cone at leaf i. Let ρ_i' and ρ_{i+1}' be the radii to leaves i and $i+1$ respectively in the plane development of the conical surface. In the transverse section, the plastochrone ratio between leaves i and $i+1$, $R_{i+1}' = \rho_i/\rho_{i+1}$. Since

$$\sin \alpha_i = \rho_i/\rho_i' = \rho_{i+1}/\rho_{i+1}'$$

(see Fig. 14), it follows that $R_{i+1}' = \rho_i'/\rho_{i+1}'$. That is, the plastochrone ratio between leaves i and $i+1$ in the plane development of the conical surface is the same as the plastochrone ratio in the transverse section.

Let θ_{i+1} be the angle between the radii to leaves i and $i+1$ in the transverse section. It subtends on the circle through i an arc of length $\rho_i\theta_{i+1}$. In the plane development of the conical surface, this arc length subtends an angle of $\theta_{i+1}' = \rho_i\theta_{i+1}/\rho_i'$. If d_{i+1} and d_{i+1}' are the divergences (fractions of a turn) that correspond to angles θ_{i+1} and θ_{i+1}' respectively, then $d_{i+1} = \theta_{i+1}/(2\pi)$, and $d_{i+1}' = \theta_{i+1}'/(2\pi)$. Consequently $d_{i+1}' = d_{i+1} \sin \alpha_i$. That is, the plane development of the conical surface shrinks the divergence between leaves i and $i+1$ by a factor equal to $\sin \alpha_i$. This shrinkage corresponds to the fact that the plane development of the conical surface transforms the surface into a sector of a circle whose angle is only $2\pi \sin \alpha_i$.

Let us now restrict our attention to the zone between leaves i and $i+1$ on the conical surface. This zone has been transformed into a sector of a ring, and the angle of the sector is $2\pi \sin \alpha_i$. The divergence between leaves i and $i+1$ can be restored to its original value d_{i+1} if the sector of the ring is stretched, like one in a fan being opened up, until its angle is 2π. The stretching factor that does this is $(\sin \alpha_i)^{-1}$. When this is done, the sector of the ring into which the zone had been transformed is converted into a full ring in a disc. But this ring is not yet a true disc representation of the zone, because it is the result of stretching the sector only tangentially. In order to preserve angles in the representation, it is necessary to stretch the ring radially, too, by the factor $(\sin \alpha_i)^{-1}$.

To see how this will affect the plastochrone ratio, let us designate the width of the ring before stretching by w, and after stretching, by $w(\sin \alpha_i)^{-1}$. Then $\rho_i' = \rho_{i+1}' + w$. Stretching the width of the ring away from the center leaves ρ_{i+1}' unchanged, but replaces ρ_i' by a new length

$$\rho_i'' = \rho_{i+1}' + w(\sin \alpha_i)^{-1}.$$

I. ADLER

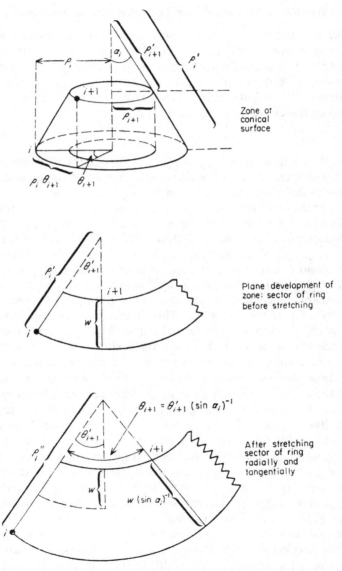

FIG. 14. Transformation of a zone of a conical surface into a zone of a disc.

For the plastochrone ratio of the ring after stretching we have

$$R_{i+1} = \rho_i''/\rho_{i+1}' = [\rho_{i+1}' + w(\sin \alpha_i)^{-1}]/\rho_{i+1}'.$$

Since

$$\rho_i' - \rho_{i+1}' = w$$

and

$$\rho_i'' - \rho_{i+1}' = w(\sin \alpha_i)^{-1},$$
$$\rho_i'' - \rho_{i+1}' = (\rho_i' - \rho_{i+1}')(\sin \alpha_i)^{-1}. \tag{C1}$$

For the plastochrone ratio before stretching we have

$$R_{i+1}' = \rho_i'/\rho_{i+1}'.$$

Then from equation (C1) we get

$$(R_{i+1} - 1) = (R_{i+1}' - 1)(\sin \alpha_i)^{-1}. \tag{C2}$$

Since R_{i+1} and R_{i+1}' are numbers of the form $1+x$ with x small,

$$R_{i+1} - 1 \doteq \ln R_{i+1}; \qquad R_{i+1}' - 1 \doteq \ln R_{i+1}'. \tag{C3}$$

Consequently,

$$\ln R_{i+1} \doteq (\sin \alpha_i)^{-1} \ln R_{i+1}'. \tag{C4}$$

APPENDIX D

Proof of Theorem 2

The proof of theorem 2 is based on combining theorem 2 of Adler (1974) with an elementary proposition about the meaning of a simple continued fraction. We begin by stating, explaining, and proving this proposition.

Proposition A

A simple continued fraction is a mediant nest.

(1) *The convergents of a simple continued fraction.* Let $a_0 + 1/a_1 + \ldots + 1/a_i + \ldots$ be a simple continued fraction. Let p_i/q_i, $i \geq 0$, be its principal convergents. Then, as is well known, p_i and q_i may be calculated by the recurrence relation

$$p_i = p_{i-2} + a_i p_{i-1}, \qquad q_i = q_{i-2} + a_i q_{i-1}, \quad i \geq 0, \tag{D1}$$

with $p_{-2} = 0$, $q_{-2} = 1$, $p_{-1} = 1$ and $q_{-1} = 0$.

[It is useful to think of $p_{-1}/q_{-1} = 1/0$ as the "principal convergent" ∞, and $p_{-2}/q_{-2} = 0/1$ as the "principal convergent" 0, assumed to precede the principal convergent p_0/q_0. The relevance of ∞ and 0 as principal convergents for all simple continued fractions will become clear in what follows.]

The simple continued fraction may also have intermediate convergents $p_{i,\,j}/q_{i,\,j}$ obtained by replacing a_i in equation (D1) by a positive integer that is less than a_i if there are are any such integers:

$$p_{i,\,j} = p_{i-2} + a_j p_{i-1}, \qquad q_{i,\,j} = q_{i-2} + a_j q_{i-1}, \quad 1 \le a_j < a_i. \qquad \text{(D2)}$$

The proposition we have set out to prove is implicit in equation (D1), but is not well known. We proceed to explain its meaning and give an indication of proof.

(2) *Terminology.* For any positive integer n, let $n/0$ represent ∞. Let us designate as a "fraction" any positive rational number, or 0, or ∞, in the form a/b, where a and b are non-negative integers, and either a or b is not zero. We say the fraction is in lowest terms if $(a, b) = 1$. Thus, 0 in lowest terms is $0/1$, and ∞ in lowest terms is $1/0$.

If inequality of fractions is defined in the usual way, that is, $a/b < c/d$ if $ad < bc$, it follows that $x < \infty$ for $x = 0$ or any positive rational number.

(3) *The mediant.* If a/b and c/d are fractions in lowest terms, and $a/b < c/d$, the mediant between a/b and c/d is defined as $(a+c)/(b+d)$. Note that $a/b < (a+c)/(b+d) < c/d$.

Examples. The mediant between $1/2$ and $1/3$ is $2/5$. If n is a non-negative integer, the mediant between n and ∞ is $n+1$. If n is a non-negative integer and m is a positive integer, the mediant between n and $n+1/m$ is $n+1/(m+1)$.

(4) *A mediant nest.* A mediant nest is a nest of closed intervals I_0, I_1, ..., I_n, ... defined inductively as follows: $I_0 = [0, \infty]$. For $n \ge 0$, if $I_n = [r, s]$, then $I_{n+1} = $ either $[r, m]$ or $[m, s]$, where m is the mediant between r and s.

It is easily shown that if at least one I_n for $n \ge 1$ has the form $[r, m]$, then the length of I_n approaches 0 as $n \to \infty$, so that such a mediant nest is truly a nest of intervals, and it determines a unique number x that is contained in every interval of the nest. For the case where every I_n for $n \ge 1$ has the form $[m, s]$, let us say that the nest determines and "contains" the number ∞.

(5) *Long notation for a mediant nest.* A mediant nest and the number it determines can be represented by a sequence of bits $b_1 b_2 b_3 \ldots b_i \ldots$, where, for $i > 0$, if $I_{i-1} = [r, s]$ and m is the mediant between r and s, $b_i = 0$ if $I_i = [r, m]$, and $b_i = 1$ if $I_i = [m, s]$.

Examples. $\dot{0} = 0$; $\dot{1} = \infty$; $\dot{1}\dot{0} = \tau$, the golden section; where each of these three examples is periodic, and the recurrent bits are indicated by the dots above them.

(6) *Abbreviated notation for a mediant nest.* The sequence of bits representing a mediant nest is a sequence of clusters of ones and zeros as follows:

$$\underbrace{b_1 b_2 b_3 \ldots b_i \ldots}_{} = \overbrace{1\ldots}^{a_0}\overbrace{10\ldots}^{a_1}\overbrace{01\ldots}^{a_2}1\ldots$$

where the a_i indicate the number of bits in each cluster; $0 \leq a_0 \leq \infty$; $0 < a_n \leq \infty$ for $n > 0$; and the sequence (a_i) terminates with a_n if $a_n = \infty$. As an abbreviated notation for a mediant nest and the number x that it determines we shall write $x = (a_0, a_1, \ldots)$. Then $a_0 \leq x < a_0 + 1$. The sequence (a_i) terminates if and only if x is rational or ∞. Every positive rational number is represented by exactly two terminating sequences (a_i).

Examples. $(\infty) = \dot{1} = \infty$; $(0, \infty) = \dot{0} = 0$; $(0, 2, \infty) = 00\dot{1} = 1/2$; $(0, 1, 1, \infty) = 01\dot{0} = 1/2$. In general, if $x = (a_0, \ldots, a_{n-1}, a_n, \infty)$ where $a_n > 1$, then $x = (a_0, \ldots, a_{n-1}, a_n - 1, 1, \infty)$, and vice versa.

Using the notation introduced above, we can restate as follows the proposition we aim to prove:

Proposition A. If $x = (a_0, a_1, \ldots, a_n, \ldots)$, then $x = a_0 + 1/a_1 + \ldots + 1/a_n + \ldots$ and conversely. If $x = (a_0, \ldots, a_n, \infty)$, then $x = a_0 + 1/a_1 + \ldots + 1/a_n$, and conversely.

(7) *Proof of the proposition.*
(i) The non-terminating case, $x = (a_0, a_1, \ldots, a_i, \ldots)$. Then x is irrational. Let p_i/q_i, for $i \geq 0$, be the principal convergents of $a_0 + 1/a_1 + \ldots$. Then a straightforward proof by induction establishes that for all odd $i \geq 1$,

$$I_{a_0 + \ldots + a_i} = [p_{i-1}/q_{i-1}, p_i/q_i],$$

and for all even $i \geq 0$,

$$I_{a_0 + \ldots + a_i} = [p_i/q_i, p_{i-1}/q_{i-1}].$$

Consequently the nest determined by successive pairs of consecutive principal convergents of $a_0 + 1/a_1 + \ldots + 1/a_n + \ldots$ defines the same number as the mediant nest $(a_0, a_1, \ldots, a_n, \ldots)$.

(ii) The terminating case, $x = (a_0, \ldots, a_n, \infty)$. It follows from (i) that

$$I_{a_0 + \ldots + a_{n+1}} = [p_n/q_n, p_{n+1}/q_{n+1}], \text{ or } [p_{n+1}/q_{n+1}, p_n/q_n],$$

where

$$p_{n+1}/q_{n+1} = (p_{n-1} + a_{n+1} p_n)/(q_{n-1} + a_{n+1} q_n).$$

Since

$$\lim_{a_{n+1} \to \infty} p_{n+1}/q_{n+1} = p_n/q_n,$$

it follows that

$$x = \lim_{a_{n+1} \to \infty} I_{a_0 + \ldots + a_{n+1}} = p_n/q_n = a_0 + 1/a_1 + \ldots + 1/a_n.$$

(iii) The "conversely" in the theorem follows from the fact that the mapping of the set of mediant nests into the set of simple continued fractions established in (i) and (ii) is one-to-one and onto.

(8) A mediant nest as defined in this appendix is obtained by starting with the interval $[0/1, 1/0]$, dividing it into two subintervals by inserting the mediant $1/1$ between its endpoints, selecting either the left or right subinterval, and repeating the same procedure with the subinterval selected, etc. In theorem 2 of Adler (1974), which determines for any given divergence d the opposed parastichy pairs that are visible, a similar construction was used starting with the interval $[1/(t+1), 1/t]$, where $t = 2, 3, \ldots$. To show the connection between the two propositions we first summarize the relevant concepts.

In any leaf distribution, the secondary spirals called *parastichies*, determined by joining a leaf to any other leaf, occur in parallel sets that effect a partition of the entire leaf distribution, so that every leaf lies on one and only one member of the set. A complete set of left parastichies (going up to the left) with m members and a complete set of right parastichies (going up to the right) with n members is called the *opposed parastichy pair* (m, n).

In an arbitrary opposed parastichy pair there need not be a leaf at every point of intersection of a left parastichy and a right parastichy. In the special case where there is a leaf at every intersection, the opposed parastichy pair is called *visible*.

If (m, n) is a visible opposed parastichy pair, there is exactly one fundamental spiral if and only if m and n are relatively prime.

If (m, n) is a visible opposed parastichy pair, with $m > n$, there is a visible opposed parastichy pair $(m-n, n)$. If $n > m$, there is a visible opposed parastichy pair $(m, n-m)$. $(m-n, n)$ or $(m, n-m)$, whichever the case may be, is called the *contraction* of (m, n).

If (m, n) is a visible opposed parastichy pair, the parastichy pair $(m+n, n)$ is called its *left extension*, and the parastichy pair $(m, m+n)$ is called its *right extension*. An extension of a visible opposed parastichy pair need not be a visible opposed parastichy pair. This fact leads to the problem of determining the conditions under which a left or right extension of a visible opposed parastichy pair is a visible opposed parastichy pair.

To solve this problem, it is shown first that, if the fundamental spiral is a right spiral, every visible opposed parastichy pair (m, n) with $m, n > 1$ can be obtained as the end product of a sequence of extensions starting with a visible opposed parastichy pair of the form $(t, t+1)$, where t is a uniquely determined integer greater than 1. It is also shown that $(t, t+1)$ is a visible opposed parastichy pair if and only if $1/(t+1) \leq d \leq 1/t$, that is, if and only if the number d is in the closed segment $[1/(t+1), 1/t]$ on the real number line.

The mediant between $1/(t+1)$ and $1/t$ is the fraction $2/(2t+1)$. It divides the segment $[1/(t+1), 1/t]$ into two segments, a left-hand segment lying to the left of the mediant, and a right-hand segment lying to the right of the mediant. It is proved that the left extension $(2t+1, t+1)$ of $(t, t+1)$ is visible if and only if d lies in the left-hand segment $[1/(t+1), 2/(2t+1)]$, and the right extension $(t, 2t+1)$ of $(t, t+1)$ is visible if and only if d lies in the right-hand segment $[2/(2t+1), 1/t]$. Moreover, this procedure can be repeated with each of these extensions. For example, start now with the fact that $(2t+1, t+1)$ is a visible opposed parastichy pair if and only if d is in the closed segment $[1/(t+1), 2/(2t+1)]$. The mediant between the end-points of this segment is $3/(3t+2)$, and it divides the segment into two segments. The left extension $(3t+2, t+1)$ of $(2t+1, t+1)$ is a visible opposed parastichy pair if and only if d lies in the left-hand segment $[1/(t+1), 3/(3t+2)]$, and the right extension $(2t+1, 3t+2)$ is a visible opposed parastichy pair if and only if d lies in the right-hand segment $[3/(3t+2), 2/(2t+1)]$ and so on.

The end product of k successive extensions starting with $(t, t+1)$ is called an extension of $(t, t+1)$ of order k. The left extension and the right extension of $(t, t+1)$ are each of order 1. Let us designate them as L and R respectively. The left extension L has both a left and right extension which are designated LL and LR respectively.

The right extension R has both a left and right extension labeled RL and RR respectively. Thus there are four extensions of $(t, t+1)$ of order 2, and their designations LL, LR, RL and RR, when read from left to right tell us how they are obtained from $(t, t+1)$. For example, LR is obtained by taking first a left extension of $(t, t+1)$, namely $(2t+1, t+1)$, and then taking a right extension of that to obtain $(2t+1, 3t+2)$. Similarly, there are eight extensions of order 3, designated as LLL, LLR, etc.

We introduce an analogous notation for successive subdivisions of the segment $[1/(t+1), 1/t]$ obtained by repeatedly inserting mediants. Inserting $2/(2t+1)$, the mediant between the endpoints of the segment, gives two segments designated as L and R: L for the left-hand segment and R for the right-hand segment. This gives the mediant subdivision of $[1/(t+1), 1/t]$ of order 1. Now by inserting in each of these segments the mediant between its endpoints, the mediant subdivision of $[1/(t+1), 1/t]$ of order 2 is obtained. It consists of four segments designated as LL, LR, RL and RR respectively. The designation, when read from left to right, indicates how the segment is obtained from $[1/(t+1), 1/t]$. The segment LR is obtained by taking the left segment of $[1/(t+1), 1/t]$, namely $[1/(t+1), 2/(2t+1)]$ and then taking the right segment of that, namely $[3/(3t+2), 2/(2t+1)]$.

The essence of theorem 2 of Adler (1974) is as follows. Extension LR of $(t, t+1)$ is a visible opposed parastichy pair if and only if the divergence d lies in segment LR of $[1/(t+1), 1/t]$; extension LL is visible if and only if d lies in segment LL; extension RL is visible if and only if d lies in segment RL; and extension RR is visible if and only if d lies in segment RR. Similarly with extensions of order 3 and the mediant subdivision of order 3, extension LRL is visible if and only if d lies in segment LRL, etc. In general, an extension of $(t, t+1)$ of order k, represented by a sequence of k letters each of which is L or R, is visible if and only if d lies in that interval of the mediant subdivision of $[1/(t+1), 1/t]$ of order k that is represented by the same sequence of letters.

(9.) *Application of proposition* A *to phyllotaxis*
When the divergence d of a leaf distribution is expressed as a simple continued fraction, proposition A makes it possible to determine all the visible opposed parastichy pairs and all the points of close return of the leaf distribution by inspection of the continued fraction. To show the procedure for doing so, we give some examples.

Example (1). $d = 5/12 = 0 + 1/2 + 1/2 + 1/2$ when expressed as a simple continued fraction. Since $a_0 = 0$ and $a_1 = 2$ (the first two terms of the continued fraction), then $1/3 \leq d \leq 1/2$. Consequently, $(2, 3)$ and its contractions $(2, 1)$ and $(1, 1)$ are visible opposed parastichy pairs. To determine the extensions of $(2, 3)$ that are visible, we proceed as follows. (a) by proposition A, $d = (0, 2, 2, 2, \infty) = 001100\dot{1}$. The first three terms 001 represent the first three intervals of the mediant nest representing d, as follows: the first 0 represents the left interval of $(0/1, 1/0)$ when the mediant $1/1$ is inserted, namely $(0/1, 1/1)$; the second 0 represents the left interval of $(0/1, 1/1)$ when the mediant $1/2$ is inserted, namely $(0/1, 1/2)$; the 1 that follows these two zeros represents the right interval of $(0/1, 1/2)$ when the mediant $1/3$ is inserted, namely $(1/3, 1/2)$. Thus the first three terms represent the fact that $1/3 \leq d \leq 1/2$. The subsequent terms $100\dot{1}$ represent the remaining intervals of the mediant nest, with each 1 representing a right interval, and each 0 representing a left interval. Thus $100\dot{1}$ is equivalent to the symbol $RLL\dot{R}$. Therefore, by theorem 2 of Adler (1974), visible extensions of $(2, 3)$ of higher and higher order are obtained by taking first a right extension, then two left extensions in succession, and then only right extensions after that. Thus we get this sequence of visible extensions of $(2,3)$:

$$\overbrace{\text{right extensions only}}$$

$(2, 5), \ (7, 5), \ (12, 5), \ (12, 17), \qquad \cdot \quad \cdot \quad \cdot$

(b) Because d is rational, we also have

$$d = (0, 2, 2, 1, 1, \infty) = 001101\dot{0}.$$

Again the first three terms 001 express the fact that $1/3 \leq d \leq 1/2$. The remaining terms $101\dot{0}$ are equivalent to the symbol $RLR\dot{L}$, and, visible extensions of $(2, 3)$ are obtained by taking in succession a right extension, a left extension, a right extension, and then only left extensions. Thus we get this sequence of visible extensions of $(2, 3)$:

left extensions only

(2, 5), (7, 5), (7, 12), (19, 12), ...

Example (2). $d = (2)^{\frac{1}{2}}/4 = 0 + 1/2 + [1/1 + 1/4]$, where the bracketed terms represent the periodic part of the non-terminating simple continued fraction that represents d. As in example (1), since $1/3 \leq d \leq 1/2$, (2, 3) and its contractions (2,1) and (1, 1) are visible.

By proposition A, $d = (0, 2, \dot{1}, \dot{4}) = 00\dot{1}0\dot{0}0\dot{0}$, which may also be written as $001\dot{0}000\dot{1}$. The first three terms 001 express the fact that $1/3 \leq d \leq 1/2$. The remaining terms $\dot{0}000\dot{1}$ are equivalent to the symbol $\dot{L}L\dot{L}L\dot{R}$, and express the fact that visible extensions of $(2, 3)$ are obtained by taking in sequence four left extensions and one right extension, and then repeating the sequence *ad infinitum*. Then the visible extensions of $(2, 3)$ are

(5, 3), (8, 3), (11, 3), (14, 3), (14, 17), ...

Example (3). $d = (3)^{\frac{1}{2}}/6 = 0 + 1/3 + [1/2 + 1/6]$. Since $1/4 \leq d \leq 1/3$, then (3, 4) and its contractions (3, 1), (2, 1) and (1, 1) are visible.

$d = (0, 3, \dot{2}, \dot{6}) = 000\dot{1}\dot{1}000000\dot{0}$, which may also be written as $0001\dot{1}0000000\dot{1}$. The first four terms 0001 express the fact that $1/4 \leq d \leq 1/3$. The remaining terms $\dot{1}0000000\dot{1}$ are equivalent to the symbol $\dot{R}L\dot{L}L\dot{L}L\dot{R}$, and express the fact that visible extensions of $(3, 4)$ are obtained by taking in sequence one right extension, six left extensions, and one right extension, and then repeating the sequence *ad infinitum*. Then the visible extensions of $(3, 4)$ are

(3, 7), (10,7), (17, 7), (24, 7), (31, 7), (38, 7), (45, 7), (45, 52), ...

Example (4). (A simplified procedure.) In the examples above, for a divergence d represented by a simple continued fraction $0 + 1/t + \ldots$ we determined separately those visible opposed parastichy pairs that are obtained by successive contractions starting with $(t, t+1)$, and those that are obtained by successive extensions starting with $(t, t+1)$. However, the two procedures can be combined into one as follows. Using the long notation for a mediant nest, write the symbol for the mediant nest that the continued fraction for d represents. Then, guided by the successive bits in the symbol, taken in order from left to right, and starting with the pseudo-parastichy-pair $(0, 1)$, take in succession a left or right extension according as the corresponding bit is 0 or 1. The first $t+1$ bits, which are $0 \ldots 01$, will yield in reverse order the visible opposed parastichy pair $(t, t+1)$ and those obtainable from it by successive contractions. All the rest will yield the successive extensions of $(t, t+1)$ of higher and higher order. For example, if d is represented by the simple continued fraction $0+1/4+1/2+1/1+\ldots$, the corresponding mediant nest is $(0, 4, 2, 1, \ldots) = 0000110 \ldots$. Then the visible opposed parastichy pairs are obtained as extensions of $(0, 1)$ by taking in succession 4 left extensions, 2 right extensions, 1 left extension, etc., to yield

$$(1, 1), \ (2, 1), \ (3, 1), \ (4, 1), \ (4, 5), \ (4, 9), \ (13, 9), \ \ldots.$$

The procedure for constructing successive visible opposed parastichy pairs establishes a one-to-one correspondence between them and the successive convergents (intermediate and principal) of the simple continued fraction for d. We tabulate this correspondence in Table 3 for example (4).

TABLE 3

i	a_i	Direction of extension	Opposed parastichy pair	Convergent
			$(0, 1)$	$p_{-1}/q_{-1} = 1/0$
				$p_0/q_0 = 0/1$
1	4	Left	$(1, 1)$	$p_{1,1}/q_{1,1} = 1/1$
			$(2, 1)$	$p_{1,2}/q_{1,2} = 1/2$
			$(3, 1)$	$p_{1,3}/q_{1,3} = 1/3$
			$(4, 1)$	$p_1/q_1 = 1/4$
2	2	Right	$(4, 5)$	$p_{2,1}/q_{2,1} = 1/5$
			$(4, 9)$	$p_2/q_2 = 2/9$
3	1	Left	$(13, 9)$	$p_3/q_3 = 3/13$

There are two properties of the visible extensions of $(0, 1)$ that are made immediately obvious in this tabulation. They are valid for the visible opposed parastichy pairs of any arbitrary divergence, so we state them as general propositions:

Proposition B

If (m, n) is a visible opposed parastichy pair for a leaf distribution with divergence d, the smaller of the two leaf numbers m and n is always the denominator of a principal convergent of the simple continued fraction for d.

Proposition C

If (m, n) is a visible opposed parastichy pair for a leaf distribution with divergence d, the larger of the two leaf numbers m and n is the denominator of a principal convergent of the simple continued fraction for d if and only if, in the sequence of visible extensions of $(0, 1)$, (m, n) immediately precedes a change in direction of extension. That is, if (m, n) is a right (left) extension of its contraction, then the larger of the two leaf numbers m and n is the denominator of a principal convergent if and only if the visible extension of (m, n) is a left (right) extension.

Since a point of close return is necessarily the denominator of a principal convergent, proposition B tells us that the smaller of the two leaf numbers m and n of a visible opposed parastichy pair is always a point of close return. Proposition C tells us when the larger of the two numbers is also a point of close return.

In example 4, the simple continued fraction for the divergence of a leaf distribution was given, and we determined all visible opposed parastichy pairs. It is also possible to reverse this procedure: Given a visible opposed parastichy pair of a leaf distribution, we can determine the range of possible values of the divergence, expressed as a simple continued fraction.

Example 5. Suppose $(17, 39)$ is a visible opposed parastichy pair. Find the range of possible values of d. Starting with $(17, 39)$, and forming successive contractions until the pseudo-opposed-parastichy pair $(0, 1)$ is reached, we obtain the following sequence of contractions:

$$
\begin{array}{l}
(17, 39) \\
\left.\begin{array}{l}(17, 22) \\ (17, 5)\end{array}\right\} 2 \text{ right contractions} \\
\left.\begin{array}{l}(12, 5) \\ (7, 5) \\ (2, 5)\end{array}\right\} 3 \text{ left contractions} \\
\left.\begin{array}{l}(2, 3) \\ (2, 1)\end{array}\right\} 2 \text{ right contractions} \\
\left.\begin{array}{l}(1, 1) \\ (0, 1)\end{array}\right\} 2 \text{ left contractions}
\end{array}
$$

If we read this tabulation in reverse order, we see how (17, 39) is obtained by successive left or right extensions starting with (0, 1):

$$
\begin{array}{l}
(0,\ 1) \\
\left.\begin{array}{l}(1,\ 1) \\ (2,\ 1)\end{array}\right\} \text{2 left extensions} \\
\left.\begin{array}{l}(2,\ 3) \\ (2,\ 5)\end{array}\right\} \text{2 right extensions} \\
\left.\begin{array}{l}(7,\ 5) \\ (12,\ 5) \\ (17,\ 5)\end{array}\right\} \text{3 left extensions} \\
\left.\begin{array}{l}(17,\ 22) \\ (17,\ 39)\end{array}\right\} \text{2 right extensions}
\end{array}
$$

Then $d = 0 + 1/2 + 1/2 + 1/3 + 1/(2+x)$, since $a_0 = 0$, $a_1 = 2$, $a_2 = 2$, $a_3 = 3$ and $a_4 \geq 2$. The possible values of x and a_4 depend on whether or not the visible extension of (17, 39) is a left extension or a right extension. If it is a left extension, $x < 1$, and $a_4 = 2$. If it is a right extension, $x \geq 1$, and $a_4 \geq 3$. Let us assume that the visible extension of (17, 39) is a left extension, so that $x < 1$. Those visible opposed parastichy pairs in the table above that immediately precede a change in direction of extension are (2, 1), (2, 5), (17, 5) and (17, 39). Since (2, 1) is the result of 2 successive left extensions of (0, 1), $(2, 1) = [0+2(1), 1]$. Since (2, 5) is the result of 2 successive right extensions of (2, 1), $(2, 5) = [2, 1+2(2)]$. Since (17, 5) is the result of 3 successive left extensions of (2, 5), $(17, 5) = [2+3(5), 5]$. Since (17, 39) is the result of 2 successive right extensions of (17, 5), $(17, 39) = [17, 5+2(17)]$. Now $(0, 1) = (q_{-1}, q_0)$, where q_{-1} and q_0 are the denominators of the principal convergents $p_{-1}/q_{-1} = 1/0$ and $p_0/q_0 = 0/1$, respectively. Then

$$
\begin{aligned}
(2,\ 1) &= (q_{-1} + a_1 q_0,\ q_0) = (q_1,\ q_0); \\
(2,\ 5) &= (q_1,\ q_0 + a_2 q_1) = (q_1,\ q_2); \\
(17,\ 5) &= (q_1 + a_3 q_2,\ q_2) = (q_3,\ q_2); \\
(17,\ 39) &= (q_3,\ q_2 + a_4 q_3) = (q_3,\ q_4).
\end{aligned}
$$

That is, if the visible opposed parastichy pair (17, 39) immediately precedes a change in the direction of extension, then the parastichy numbers 17 and 39 are necessarily denominators of *consecutive* principal convergents of the simple continued fraction for d. This result can be generalized as follows:

Proposition D

Let either (p, q) or (q, p) be a visible opposed parastichy pair which immediately precedes a change in direction of visible extension. That is, if (p, q) or (q, p) is a right extension of its contraction, then only the left extension of (p, q) or (q, p) is visible, and if (p, q) or (q, p) is a left extension of its contraction, then only the right extension of (p, q) or (q, p) is visible. Then p and q are necessarily denominators of *consecutive* principal convergents of the simple continued fraction for the divergence of the leaf distribution.

Proof. By successive contractions until we reach $(0, 1)$, we identify integers $a_1, \ldots, a_n$ such that (p, q) or (q, p) is the end result of successive extensions of $(0, 1)$ as follows:

a_1 left extensions followed by

a_2 right extensions followed by

$\cdot \ \cdot \ \cdot$

a_n left or right extensions
according as n is odd or even.

Then $d = 0 + 1/a_1 + 1/a_2 + \ldots + 1/(a_n + x)$, where $x < 1$. But $(0, 1) = (q_{-1}, q_0)$ where q_{-1} and q_0 are the denominators of the principal convergents $1/0$ and $0/1$ respectively. The result of a_1 left extensions of (q_{-1}, q_0) is

$$(q_{-1} + a_1 q_0, q_0) = (q_1, q_0).$$

The result of a_2 right extensions of (q_1, q_0) is

$$(q_1, q_0 + a_2 q_1) = (q_1, q_2),$$

etc.

Theorem 2 of section 6 then follows as a corollary of propositions B, C and D.

APPENDIX E

Computed Predicted Values of d(T) for Given Monotonic Decreasing Functions r(T)

TABLE 4

$r = 1{\cdot}5T^{-2}$; $T_c = 3$; initial d is between 1/3 and 1/2

T	d	Phyllotaxis	T	d	Phyllotaxis
3	0·3780	(1, 2)	14	0·3834	(5, 8)
4	0·3767	(2, 3)	15	0·3837	(8, 13)
5	0·3908	(2, 3)	16	0·3830	(8, 13)
			17	0·3825	(8, 13)
6	0·3909	(3, 5)	18	0·3822	(8, 13)
7	0·3830	(3, 5)	19	0·3819	(8, 13)
8	0·3796	(3, 5)	20	0·3817	(8, 13)
9	0·3778	(3, 5)	21	0·3816	(8, 13)
			22	0·3815	(8, 13)
10	0·3798	(5, 8)	23	0·3814	(8, 13)
11	0·3814	(5, 8)	24	0·3813	(8, 13)
12	0·3824	(5, 8)			
13	0·3830	(5, 8)	25	0·3815	(13, 21)

d converges to $1/2 + 1/\tau = 0{\cdot}381966$.

TABLE 5

$r = 1{\cdot}5T^{-2}$; $T_c = 6$; initial d is between 2/5 and 1/2

T	d	Phyllotaxis	T	d	Phyllotaxis
6	0·4040	(2, 5)	19	0·4202	(7, 12)
7	0·4174	(2, 5)	20	0·4204	(7, 12)
8	0·4224	(2, 5)	21	0·4205	(7, 12)
9	0·4211	(5, 7)	22	0·4206	(12, 19)
10	0·4195	(5, 7)	23	0·4203	(12, 19)
11	0·4186	(5, 7)	24	0·4202	(12, 19)
12	0·4180	(5, 7)	25	0·4200	(12, 19)
13	0·4176	(5, 7)	26	0·4199	(12, 19)
			28	0·4198	(12, 19)
14	0·4178	(7, 12)	30	0·4197	(12, 19)
15	0·4187	(7, 12)	32	0·4196	(12, 19)
16	0·4193	(7, 12)	34	0·4195	(12, 19)
17	0·4197	(7, 12)			
18	0·4200	(7, 12)	36	0·4195	(19, 31)

d converges to $1/2 + 1/2 + 1/\tau = 0{\cdot}419821$.

TABLE 6

$r = T^{-1}/4; T_c = 2;$ *initial* d *is between 1/3 and 1/2*

T	d	Phyllotaxis	T	d	Phyllotaxis
2	0·3577	(1, 2)	13	0·3780	(3, 5)
			14	0·3776	(3, 5)
3	0·3818	(2, 3)			
4	0·3900	(2, 3)	15	0·3785	(5, 8)
5	0·3936	(2, 3)	16	0·3793	(5, 8)
			17	0·3800	(5, 8)
			18	0·3805	(5, 8)
6	0·3909	(3, 5)	19	0·3810	(5, 8)
7	0·3862	(3, 5)	20	0·3814	(5, 8)
8	0·3834	(3, 5)	21	0·3917	(5, 8)
9	0·3815	(3, 5)	22	0·3820	(5, 8)
10	0·3802	(3, 5)	23	0·3822	(5, 8)
11	0·3793	(3, 5)	24	0·3824	(5, 8)
12	0·3786	(3, 5)	25	0·3826	(5, 8)

d converges to $1/2 + 1/\tau$.

TABLE 7

$r = T^{-0.3}/15; T_c = 2;$ *initial* d *is between 1/3 and 1/2*

T	d	Phyllotaxis	T	d	Phyllotaxis
2	0·3378	(1, 2)	12	0·3836	(3, 5)
			13	0·3832	(3, 5)
3	0·3942	(2, 3)	14	0·3828	(3, 5)
4	0·3951	(2, 3)	15	0·3824	(3, 5)
			20	0·3812	(3, 5)
5	0·3904	(3, 5)	25	0·3804	(3, 5)
6	0·3886	(3, 5)	30	0·3798	(3, 5)
7	0·3873	(3, 5)	50	0·3785	(3, 5)
8	0·3862	(3, 5)	70	0·3778	(3, 5)
9	0·3854	(3, 5)			
10	0·3847	(3, 5)	90	0·3779	(5, 8)
11	0·3841	(3, 5)	100	0·3784	(5, 8)

d converges to $1/2 + 1/\tau$.

TABLE 8

$r = (2\pi)^{-1} \ln \{(1+aT)/[1+a(T-1)]\}$, a $= 0.14$; $T_c = 2$; *initial* d *is between* 1/3 *and* 1/2; nr *is used as vertical component of* dist $(0, n)$. (*Cylindrical representation of disc model*)

T	d	Phyllotaxis	T	d	Phyllotaxis
2	0·33384	(1, 2)	34	0·38179	(8, 13)
			35	0·38175	(8, 13)
3	0·39932	(2, 3)	36	0·38171	(8, 13)
4	0·39944	(2, 3)	37	0·38168	(8, 13)
			38	0·38164	(8, 13)
5	0·37652	(3, 5)	39	0·38161	(8, 13)
6	0·37628	(3, 5)	40	0·38158	(8, 13)
7	0·37610	(3, 5)	41	0·38156	(8, 13)
			42	0·38153	(8, 13)
8	0·38220	(5, 8)	43	0·38151	(8, 13)
9	0·38251	(5, 8)	44	0·38149	(8, 13)
10	0·38276	(5, 8)	45	0·38146	(8, 13)
11	0·38298	(5, 8)	46	0·38145	(8, 13)
12	0·38315	(5, 8)	47	0·38143	(8, 13)
13	0·38330	(5, 8)	48	0·38141	(8, 13)
14	0·38343	(5, 8)	49	0·38139	(8, 13)
15	0·38354	(5, 8)	50	0·38138	(8, 13)
16	0·38363	(5, 8)			
17	0·38372	(5, 8)	56	0·38133	(13, 21)
			60	0·38147	(13, 21)
18	0·38348	(8, 13)	70	0·38171	(13, 21)
19	0·38325	(8, 13)	80	0·38186	(13, 21)
20	0·38306	(8, 13)	90	0·38196	(13, 21)
21	0·38289	(8, 13)	100	0·38204	(13, 21)
22	0·38274	(8, 13)	110	0·38209	(13, 21)
23	0·38261	(8, 13)	120	0·38213	(13, 21)
24	0·38249	(8, 13)	130	0·38216	(13, 21)
25	0·38239	(8, 13)	140	0·38219	(13, 21)
26	0·38230	(8, 13)	150	0·38221	(13, 21)
27	0·38221	(8, 13)			
28	0·38213	(8, 13)	156	0·38222	(21, 34)
29	0·38206	(8, 13)	160	0·38220	(21, 34)
30	0·38200	(8, 13)	170	0·38215	(21, 34)
31	0·38194	(8, 13)	180	0·38211	(21, 34)
32	0·38189	(8, 13)	190	0·38208	(21, 34)
33	0·38184	(8, 13)	200	0·38205	(21, 34)

d converges to $1/2 + 1/\tau$.

TABLE 9

$r = (2\pi)^{-1} \ln (1+aT)/[1+a(T-1)]$, $a = 0.14$; $r_i(T) = r(T-i+1)$; $T_c = 2$; initial d *is between 1/3 and 1/2;* $r_1 + \ldots + r_n$ *is used as vertical component of dist* (0, n). (*Cylindrical representation of disc model*)

T	d	Phyllotaxis	T	d	Phyllotaxis
2	0·33394	(1, 2)	21	0·38367	(5, 8)
			22	0·38375	(5, 8)
3	0·39905	(2, 3)	23	0·38381	(5, 8)
4	0·39925	(2, 3)	24	0·38388	(5, 8)
5	0·37760	(3, 5)			
6	0·37709	(3, 5)	25	0·38372	(8, 13)
7	0·37671	(3, 5)	30	0·38272	(8, 13)
			35	0·38220	(8, 13)
8	0·37885	(5, 8)	40	0·38188	(8, 13)
9	0·38001	(5, 8)	45	0·38167	(8, 13)
10	0·38084	(5, 8)	50	0·38153	(8, 13)
11	0·38145	(5, 8)	55	0·38142	(8, 13)
12	0·38193	(5, 8)	60	0·38134	(8, 13)
13	0·38230	(5, 8)	65	0·38128	(8, 13)
14	0·38260	(5, 8)			
15	0·38284	(5, 8)	70	0·38138	(13, 21)
16	0·38304	(5, 8)	80	0·38165	(13, 21)
17	0·38321	(5, 8)	90	0·38183	(13, 21)
18	0·38335	(5, 8)	100	0·38194	(13, 21)
19	0·38347	(5, 8)	110	0·38202	(13, 21)
20	0·38358	(5, 8)	120	0·38208	(13, 21)

d converges to $1/2 + 1/\tau$.

TABLE 10

$r = (4\pi)^{-1}[1+4a(T+b)]^{\frac{1}{2}} \, ln \, [(T+b)/(T+b-1)]$, $a = 0 \cdot 1$, $b = 7$; nr *is used as vertical component of dist* (0, n). (*Cylindrical representation of parabolic model.*) $T_c = 2$; *initial* d *is between 1/3 and 1/2*

T	d	Phyllotaxis	T	d	Phyllotaxis
2	0·3339	(1, 2)	17	0·3821	(5, 8)
			18	0·3822	(5, 8)
3	0·3991	(2, 3)	19	0·3823	(5, 8)
4	0·3992	(2, 3)	20	0·3824	(5, 8)
			30	0·3831	(5, 8)
5	0·3771	(3, 5)	40	0·3835	(5, 8)
6	0·3770	(3, 5)	50	0·3837	(5, 8)
7	0·3769	(3, 5)			
			60	0·3833	(8, 13)
8	0·3801	(5, 8)	70	0·3830	(8, 13)
9	0·3805	(5, 8)	80	0·3827	(8, 13)
10	0·3808	(5, 8)	90	0·3825	(8, 13)
11	0·3811	(5, 8)	100	0·3823	(8, 13)
12	0·3813	(5, 8)	200	0·3816	(8, 13)
13	0·3815	(5, 8)	300	0·3814	(8, 13)
14	0·3817	(5, 8)			
15	0·3818	(5, 8)			
16	0·3820	(5, 8)	400	0·3814	(13, 21)

d converges to $1/2 + 1/\tau$.

TABLE 11

$r = (4\pi)^{-1}[1+4a(T+b)]^{\frac{1}{2}} \, ln \, (T+b)/(T+b-1)$, $a = 0 \cdot 1$, $b = 7$; nr *is used as vertical component of dist* (0, n). (*Cylindrical representation of parabolic model.*) $T_c = 5$; *initial* d *is between 2/5 and 1/2*

T	d	Phyllotaxis	T	d	Phyllotaxis
5	0·4256	(2, 5)	19	0·4180	(5, 7)
6	0·4259	(2, 5)	20	0·4180	(5, 7)
			30	0·4176	(5, 7)
7	0·4195	(5, 7)			
8	0·4193	(5, 7)	40	0·4178	(7, 12)
9	0·4191	(5, 7)	50	0·4185	(7, 12)
10	0·4189	(5, 7)	60	0·4190	(7, 12)
11	0·4188	(5, 7)	70	0·4193	(7, 12)
12	0·4186	(5, 7)	80	0·4195	(7, 12)
13	0·4185	(5, 7)	90	0·4197	(7, 12)
14	0·4184	(5, 7)	100	0·4198	(7, 12)
15	0·4183	(5, 7)	200	0·4204	(7, 12)
16	0·4182	(5, 7)			
17	0·4182	(5, 7)	300	0·4204	(12, 19)
18	0·4181	(5, 7)	400	0·4201	(12, 19)

d converges to $1/2 + 1/2 + 1/\tau$.

APPENDIX F

Proof that Contact Pressure Compels Equalization of Divergences

Let p and q be the leaves nearest zero such that (d, r) is on the (p, q) semicircle, with p and q on opposite sides of l_0. p and q are relatively prime, since we are assuming that there is one fundamental spiral. The condition $[d, r(T)]$ is on the (p, q) semicircle determines d as a function of r such that $d_1 + \ldots + d_p = pd$ and $d_1 + \ldots + d_q = qd$. (See section 1, paragraph 8, note 2.) Since any leaf n may play the role of leaf zero, we have, in general,

$$d_n + \ldots + d_{n+p-1} = pd, \qquad (F1)$$

$$d_n + \ldots + d_{n+q-1} = qd. \qquad (F2)$$

Replacing n by $n+1$ in equation (F1), we get

$$d_{n+1} + \ldots + d_{n+p} = pd. \qquad (F3)$$

It follows from equations (F1) and (F3) that

$$d_n = d_{n+p}. \qquad (F4)$$

Similarly,

$$d_n = d_{n+q}. \qquad (F5)$$

Therefore, $d_m = d_n$ if $m \equiv n$ modulo p, and $d_m = d_n$ if $m \equiv n$ modulo q.

For arbitrary positive integers m and n we show that there exists an integer k such that $m \equiv k$ modulo p, and $n \equiv k$ modulo q. (This is a special case of the Chinese remainder theorem.) It will follow from this that $d_m = d_k = d_n$.

Let r_m be the least residue of m modulo p. Let r_n be the least residue of n modulo q. Then $m \equiv r_m$ modulo p, and $n \equiv r_n$ modulo q. Since p and q are relative prime, there exists an integer a with $1 \leq a \leq q$ such that $pa \equiv r_n - r_m$ modulo q. Then $pa + r_m \equiv r_n$ modulo q. But $n \equiv r_n$ modulo q. Therefore

$$n \equiv pa + r_m \text{ modulo } q. \qquad (F6)$$

Obviously $r_m \equiv pa + r_m$ modulo p. But $m \equiv r_m$ modulo p. Therefore,

$$m \equiv pa + r_m \text{ modulo } p. \qquad (F7)$$

Then $pa + r_m$ is the desired number k.

Thus as T increases, more and more d_i are equalized at the value of d determined by the condition $[d, r(T)]$ is on the (p, q) semicircle. This continues until r becomes small enough for leaf $p+q$ to displace either p or q, whichever has the smaller leaf number, as leaf nearest zero. Then (if p is the smaller leaf number), the process of equalization of the d_i starts all over again at the value of d determined by the condition that $[d, r(T)]$ is on the $(q, p+q)$ semicircle.

J. theor. Biol. (1977) **67**, 447–458

An Application of the Contact Pressure Model of Phyllotaxis to the Close Packing of Spheres around a Cylinder in Biological Fine Structure

IRVING ADLER

North Bennington, Vermont 05257, U.S.A.

(*Received* 21 *December* 1976)

Equations derived from a contact pressure model of phyllotaxis are relevant to close packing of equal spheres on a cylindrical surface. They provide a general method of calculating the parameters for microscopic biological structures that are assembled in helical arrangements of protein monomers. In the triple-contact case of hexagonal packing, with km, kn and $k(m+n)$ contact parastichies respectively, where m and n are relatively prime, the divergence angle d and the normalized internode distance r on the k fundamental spirals are completely determined by m, n and k, and formulas are given for calculating them. In the double-contact case of rhombic packing, with km and kn contact parastichies respectively, where $m < n$, d is a function of r, and the domain of r is the interval between the value of r determined for the triple-contact case of km, kn and $k(m+n)$ parastichies and the value of r determined for the triple-contact case of $k(n-m)$, km and kn parastichies. Here r and d can be determined from the measured ratio of the radius of the cylindrical surface to the radius of the spheres.

1. Phyllotaxis and Some Biological Fine Structure

Erickson (1973) proposed the use of the concepts and terminology of phyllotaxis to describe microscopic biological structures that are assembled from protein monomers in helical arrangements like those displayed by the close packing of equal spheres around a cylindrical surface. To calculate appropriate phyllotaxis parameters from observed data, he used two methods employing the notation and the equations of Van Iterson (1907). Method 1 (Erickson, 1973, pp. 708–9) is based on using for the distance between the centers of two neighboring spheres the length of the straight line-segment between them. This method leads to transcendental equations that can be solved by an iterative procedure. Method 2 (Erickson, 1973, p. 710) uses for the distance between the centers of two neighboring spheres the length of the helical arc that joins them on the cylindrical surface which contains the centers. This method leads to simple algebraic equations that are solved more

142

directly than the transcendental equations of Method 1 and provides solutions that are close approximations to those obtained by Method 1.

The equations Erickson used for Method 2 are applicable only to the triple-contact case (hexagonal packing). This paper develops a generalization of Method 2 applicable to both the triple-contact case and the double-contact case (rhombic packing). The generalization is based on the equations derived in Adler (1974, 1977) for a model of contact pressure in phyllotaxis.

For the derivation of the Van Iterson equations, parameters are normalized by taking as unit of length the radius of the closely packed spheres. The Adler equations employ a different normalization in which the unit of length is the girth of the cylindrical surface through the centers of the spheres. The Adler normalization leads to a simplification of the results and reveals relationships that are obscured by the other notation. [For example, for any given rhombic packing displaying m right spirals and n left spirals, throughout the range of possible values of the divergence angle, the divergence angle is a linear function of the square of the diameter of the spheres. [See equation (18) in section 10 below.] A table for translating from one notation to the other is provided below after the necessary definitions are given.

2. Jugacy of a Cylindrical Point-lattice

Consider a uniform distribution of closely packed equal spheres around a cylindrical surface in which the centers of the spheres also lie on a cylindrical surface and constitute a point-lattice on the surface. It is this cylindrical point-lattice that concerns us in the remainder of this paper.

The cylindrical point-lattice becomes a plane point-lattice in the plane development of the cylindrical surface. To picture the contact relations of the closely-packed spheres whose centers are the lattice points on the cylindrical surface, in the plane development of this surface we replace the spheres by closely packed circles whose centers are the lattice points in the plane. Let δ be the diameter of each of these closely-packed circles. It is a close approximation to the diameter of the closely-packed spheres.

(The closely-packed circles in the plane may also be used to represent approximately the contact relations of some other packings of equal spheres around a cylinder in which the centers of the spheres are not all on the same cylindrical surface. An example of such a packing is one in which black and white spheres are arranged in a checkerboard pattern coiled around a cylinder, with the centers of the black spheres nearer the axis than those of the white spheres. Then, in this case, the centers of the black spheres are on one cylindrical surface, and the centers of the white spheres are on another cylindrical surface with the same axis. This packing can be represented by a cylindrical point lattice obtained by projecting the centers of all the spheres radially from the axis of the cylinders onto one cylindrical surface. In this representation, however, the diameters of the closely packed circles drawn around the lattice points will be somewhat smaller than the diameters of the spheres they represent.)

We shall take the axis of the cylinder to be vertical, and we shall view the cylinder from the outside, in order to give unambiguous meanings to the terms *higher, left, right, vertical,* and *level.* (The noun *level* will refer to a horizontal plane.)

For any given lattice point there may or may not be other lattice points at the same level. If the total number of lattice points at the level of a given lattice point is $k \geq 1$, the lattice is said to be *k-jugate.* If $k > 1$, a 1-jugate lattice intimately related to the given lattice can be formed in a simple way. Slit the cylindrical surface along the element through each of the k lattice points that are on a given level, thus cutting the cylindrical surface into k strips. Roll up one of these strips to form a cylindrical surface in which the two vertical edges of the strip coincide. Then the lattice points on the strip become a 1-jugate cylindrical point-lattice.

In the derivation of the equations for calculating the parameters of a cylindrical point-lattice, we shall deal first with 1-jugate lattices. For $k > 1$, the parameters of a k-jugate lattice are derived from the parameters of the related 1-jugate lattice.

3. Fundamental Spirals

In a 1-jugate cylindrical lattice, the helix that joins each lattice point to the next higher lattice point (the short way around) is called the *fundamental spiral.* We number the lattice points starting with 0, in order of height along the fundamental spiral: 0, 1, 2, 3, ...

In a k-jugate cylindrical lattice there are k parallel fundamental spirals, one through each of the k lattice points at each level that contains any.

We normalize the lattice by taking the girth of the cylindrical surface to be 1. For a given jugacy k, the state of the lattice is completely determined by two parameters: $d = divergence = $ the horizontal component of the distance between two consecutive lattice points = the angle between them around the axis of the cylinder measured as a fraction of a turn the short way around; $r = rise = $ the normalized internode distance = the vertical component of the distance between two consecutive lattice points. The state of the lattice may be represented as a point in a two-dimensional phase space, namely the (d, r) plane.

Let (d, r) be the parameters for a k-jugate point-lattice. To form the related

1-jugate lattice we roll into a cylinder a strip whose width is $1/k$, thus transforming into 360° what was an angle of $360°/k$. The divergence d in the k-jugate lattice is transformed into a divergence d' in the 1-jugate lattice with

$$d' = kd. \tag{1}$$

To find the rise r' between consecutive lattice points in the 1-jugate lattice, we must renormalize the vertical distance r, using as unit of length the girth $1/k$ of the cylindrical surface on which the 1-jugate lattice lies. Then

$$r' = kr. \tag{2}$$

4. Comparison of Notations

In Erickson (1973) the following notations are used: The radius of the packed spheres is taken as unit of length; the divergence angle is called α; the radius of the cylinder is called R; and the internode distance is called h. The relationship between Erickson's notation and the notation of this paper is summarized in the table below.

TABLE 1

Magnitude	Erickson's notation	Transformation $\rightarrow$	Notation of this paper
Girth of cylinder	$2\pi R$	$\div 2\pi R =$	1
Internode distance	h	$\div 2\pi R =$	r
Diameter of sphere	2	$\div 2\pi R =$	δ
Divergence	α	$=$	d

Consequently, we have

$$\pi R \delta = 1, \tag{3}$$

$$h = 2\pi R r. \tag{4}$$

Note that if measurements are made using an arbitrary unit of length, the (Erickson) normalized measure R is obtained as follows:

$$R = (\text{radius of cylinder})/(\text{radius of sphere}). \tag{5}$$

5. Contact Parastichies

The spirals that are conspicuous in a uniform distribution of closely packed equal spheres are referred to in the botanical literature as *contact parastichies*. They are the spirals which join each lattice point to its nearest neighbors.

If the lattice point nearest to lattice point 0 and on its right (left) is lattice point m, then the parastichy through 0 and m, and the parastichies parallel to it, form a set of exactly m parastichies going up to the right (left) (Adler, 1974). Thus the conspicuous parastichies occur in parallel sets. Under conditions of close packing, there may be two or three such sets. (That no more than three sets are possible follows from the fact that in a closely-packed array of equal circles in a plane whose centers are the lattice points, each circle can be tangent to at most six circles surrounding it, and each pair of the surrounding circles that are on opposite sides of the central circle determine one parastichy set.)

If there are two sets (double contact, rhombic packing), one set goes up to the left (left parastichies) and the other set goes up to the right (right parastichies). If there are x spirals that go up in one direction, and y spirals that go up in the other, the cylindrical lattice is said to display (x, y) phyllotaxis. Let

$$k = \text{the greatest common divisor of } x \text{ and } y, \left.\right\}$$
$$m = x/k, \qquad n = y/k. \qquad\qquad\qquad \tag{6}$$

Then the point-lattice is k-jugate (Bravais & Bravais, 1837), and m and n are relatively prime. We shall write $k(m, n)$ to represent (km, kn) phyllotaxis. If $k = 1$, the k will be omitted.

If there are three sets of contact parastichies (triple contact, hexagonal packing), and the numbers of the spirals in the three sets are x, y and z respectively, then one of the three numbers, say z, is greater than the other two, and, in fact, $z = x+y$ (Adler, 1977; see section 7 below). The x spirals and the y spirals are in opposite directions. The $x+y$ spirals are in the same direction as the x spirals if $x < y$, and in the opposite direction if $x > y$. As in the case of rhombic packing, if k is the greatest common divisor of x and y, the point-lattice is k-jugate. If $m = x/k$, $n = y/k$, we shall write $k(m, n; m+n)$ to represent the type of phyllotaxis involved. It includes $k(m, n)$ phyllotaxis in the sense defined above: km spirals in one direction, and kn spirals in the other. It also includes at the same time either $k(n, m+n)$ phyllotaxis or $k(m, m+n)$ phyllotaxis, according as $m < n$ or $m > n$.

6. Observables and the Problem to be Solved

In microscopic biological structures that look like helical arrangements of closely packed equal spheres around a cylinder, the following are observables:

(1) the numbers of conspicuous spirals x, y in rhombic packing; the numbers of conspicuous spirals x, y, $x+y$ in hexagonal packing;

(2) the direction, right or left, of each set of conspicuous spirals;
(3) the radius of the cylinder;
(4) the radius of the spheres.

From these observables, the following numbers and measures are immediately determined by equations (6), (5) and (3): k, m, n, R, and δ.

Our problem is to use this information to calculate d and r.

7. The Relevance of the Contact Pressure Model

In the model of contact pressure in phyllotaxis constructed by Adler (1974, 1977), contact pressure is represented by the assumption that the minimum distance between lattice points is maximized. It is then proved that if m and n are the lattice points nearest lattice point 0, and are on opposite sides of 0, they must satisfy the condition

$$\text{dist}^2(0, m) = \text{dist}^2(0, n). \tag{7}$$

This condition defines a path in the (d, r) plane, and d and r are determined as coordinates of points on this path.

Condition (7) is automatically fulfilled in the close-packing of equal spheres around a cylinder, where $\text{dist}\,(0, m)$ and $\text{dist}\,(0, n)$ are both equal to the diameter of the spheres. Therefore the consequences of equation (7) are applicable to structures that have the form of arrays of equal spheres closely packed around a cylinder.

8. Applicable Results of the Contact Pressure Model

The following results are taken from Adler (1977).

(1) If the numbers of conspicuous left parastichies and right parastichies are m and n, not necessarily respectively, with $m < n$, then $m = q_{i-1}$, $n = q_i$, where q_{i-1} and q_i are denominators of consecutive principal convergents p_{i-1}/q_{i-1} and p_i/q_i of the simple continued fraction for the divergence d.

(2) The condition (7) requires that the point (d, r) lie on an arc of the semi-circle known as the (q_{i-1}, q_i) semicircle whose equation is

$$[d - (p_i q_i - p_{i-1} q_{i-1})/(q_i^2 - q_{i-1}^2)]^2 + r^2$$
$$= 1/(q_i^2 - q_{i-1}^2)^2, \qquad r > 0. \tag{8}$$

The arc extends from the intersection of the (q_{i-1}, q_i) semicircle with the $(q_i - q_{i-1}, q_{i-1})$ semicircle to the intersection of the (q_{i-1}, q_i) semicircle with the $(q_i, q_{i-1} + q_i)$ semicircle.

(3) On this arc,

$$\delta^2 = \text{dist}^2\,(0,\,q_{i-1}) = q_{i-1}^2 r^2 + (q_{i-1}d - p_{i-1})^2, \tag{9}$$

$$\delta^2 = \text{dist}^2\,(0,\,q_i) = q_i^2 r^2 + (q_i d - p_i)^2. \tag{10}$$

9. Applicable Properties of the Simple Continued Fraction for d

In order to use equation (8), it is necessary first to calculate the values of p_{i-1} and p_i. A method of making this calculation is outlined here based on a well-known property of the simple continued fraction for the divergence d. In the simple continued fraction

$$a_0 + \cfrac{1}{a_1 + \cfrac{1}{a_2 + \ldots}}$$

for a positive real number d, consecutive principal convergents p_i/q_i satisfy the following equation (Le Veque, 1956):

$$q_i p_{i-1} - p_i q_{i-1} = (-1)^i. \tag{11}$$

Consequently, the ordered pair $(p_{i-1},\,p_i)$ is a solution of the linear Diophantine equation

$$q_i x - q_{i-1} y = (-1)^i. \tag{12}$$

If $(x_1,\,y_1)$ is a particular solution of equation (12), all solutions are given by

$$x = x_1 + t q_{i-1},\, y = y_1 + t q_i, \tag{13}$$

where t is any integer. Since d is a divergence angle, we have $d < 1$. It follows that $0 < p_{i-1} < q_{i-1}$, and $0 < p_i < q_i$. Consequently the pair $(p_{i-1},\,p_i)$ that we seek is the solution of equation (12) that satisfies the condition

$$0 < x < q_{i-1}, \qquad 0 < y < q_i. \tag{14}$$

It is provided by the usual method (Le Veque, 1956, p. 16), based on the Euclidean algorithm, for showing that solutions of equation (12) exist:
Write $r_1 = q_i$, and $r_2 = q_{i-1}$, and divide r_1 by r_2. Let a_1 and r_3 respectively be the quotient and remainder obtained. If $r_3 > 1$, divide r_2 by r_3, and let a_2 and r_4 respectively be the quotient and remainder obtained. Continue in this way, dividing each new remainder r_i into the preceding one until a remainder

r_{n+2} is obtained that is equal to 1. Then we have this sequence of equations:

$$\left.\begin{array}{l} r_1 = a_1 r_2 + r_3, \\ r_2 = a_2 r_3 + r_4, \\ \cdots \\ r_{n-1} = a_{n-1} r_n + r_{n+1}, \\ r_n = a_n r_{n+1} + 1. \end{array}\right\} \qquad (15)$$

From the last of these equations we have $r_n - a_n r_{n+1} = 1$. The equations before the last one, taken in reverse order, give us expressions that may be substituted in succession for $r_{n+1}, r_n, r_{n-1}, \ldots, r_3$, permitting us to eliminate them one by one. The final result is an equation of the form $r_1 x - r_2 y = (-1)^i$, which provides a solution of equation (12), since $r_1 = q_i$, and $r_2 = q_{i-1}$. Moreover, it can be proved by induction on n that the x and y obtained in this way satisfy condition (14), so they are the numbers p_{i-1} and p_i respectively that we seek.

Example. Find p_{i-1} and p_i if $q_{i-1} = 3$ and $q_i = 7$. $7 = 2 \cdot 3 + 1$, or $1 \cdot 7 - 2 \cdot 3 = 1$. Therefore $p_{i-1} = 1$, and $p_i = 2$.

Example. Find p_{i-1} and p_i if $q_{i-1} = 3$ and $q_i = 17$. $17 = 5 \cdot 3 + 2$, and $3 = 1 \cdot 2 + 1$. Then

$$1 = 3 - 1 \cdot 2$$
$$1 = 3 - 1 \cdot (17 - 5 \cdot 3),$$
$$1 = 6 \cdot 3 - 1 \cdot 17.$$

Consequently $1 \cdot 17 - 6 \cdot 3 = -1$. Then $p_{i-1} = 1$, and $p_i = 6$.

10. The 1-Jugate Case, with Hexagonal Packing

In this case, the only observational data needed to calculate d, r, δ, R, and h are the numbers of conspicuous parastichies x, y, z with $x + y = z$. The calculation proceeds via the following sequence of steps.

    (1) Since we are assuming $k = 1$, the x and y are relatively prime. q_{i-1} is the smaller and q_i is the greater of these two numbers. $q_{i+1} = q_{i-1} + q_i$.

    (2) Calculate p_{i-1}, p_i, and i by the procedure described in section 9. $p_{i+1} = p_{i-1} + p_i$.

    (3) The equation of the (q_{i-1}, q_i) semicircle is equation (8) in section 8. Write the corresponding equation for the (q_i, q_{i+1}) semicircle).

(4) d and r, the coordinates of the point of intersection of these two semi-circles, have these values:

$$d = \frac{q_{i+1}p_{i-1}(q_i-q_{i-1})(p_i+p_{i+1})-q_{i-1}p_{i+1}(q_i+q_{i+1})(p_i-p_{i-1})}{2[q_{i+1}(q_i-q_{i-1})(p_{i+1}q_{i+1}-p_iq_i)+q_{i-1}(q_i+q_{i+1})(p_{i-1}q_{i-1}-p_iq_i)]} \quad (16)$$

$$r = \frac{\{1-[(q_i^2-q_{i-1}^2)d-(p_iq_i-p_{i-1}q_{i-1})]^2\}^{\frac{1}{2}}}{q_i^2-q_{i-1}^2} \quad (17)$$

(5) Eliminating d^2 and r^2 from equations (9) and (10), and invoking equation (11) give

$$d = \tfrac{1}{2}[p_i/q_i+p_{i-1}/q_{i-1}+e\delta^2(q_i/q_{i-1}-q_{i-1}/q_i)], \quad (18)$$

where $e = (-1)^{i+1}$. Use equation (18) and the value of d derived from equation (16) to calculate δ.

(6) Use equation (3) to calculate R.

(7) Use equation (4) to calculate h.

It is interesting to note the meaning of equation (18). From the properties of simple continued fractions we know that d lies between the consecutive principal convergents p_{i-1}/q_{i-1} and p_i/q_i. So it is not surprising that their average is a good approximation to d. The new information provided by equation (18) is that the difference between d and this average depends only on the ratio of the parastichy numbers q_{i-1} and q_i, the parity of i, and the area occupied by each of the closely packed equal circles. (This area is $\pi\delta^2/4$.)

11. The 1-Jugate Case with Rhombic Packing

In this case, the observational data needed to calculate d, r, δ and h are R (the ratio of the radius of the cylinder to the radius of the spheres), and the numbers of conspicuous parastichies x and y. x and y are necessarily relatively prime. The calculation proceeds in this sequence of steps.

(1) q_{i-1} is the smaller, and q_i is the greater of the two numbers x and y.

(2) Calculate p_{i-1} and p_i by the procedure described in section 9.

(3) Let $q_{i+1} = q_i+q_{i-1}$, $q_{i-2} = q_i-q_{i-1}$, $p_{i+1} = p_i+p_{i-1}$, $p_{i-2} = p_i-p_{i-1}$.

(4) The range of possible values of d is the range between the value of d given by equation (16) and the value given by (16) when i is replaced by $i-1$.

(5) The range of possible values of r is the range between the value of r given by equation (17) and the value given by (17) when i is replaced by $i-1$.

(6) Use equation (3) to calculate δ.

I. ADLER

(7) Use equation (18) to calculate d.

(8) Use equation (8) to calculate r.

(9) Use equation (4) to calculate h.

12. The k-Jugate Cases

If the observed contact phyllotaxis is $k(m, n)$, $k > 1$, with rhombic packing, or $k(m, n; m+n)$, $k > 1$, with hexagonal packing, the calculations for d and r and related parameters are made as follows:

(1) q_{i-1} is the smaller, and q_i is the greater of the two numbers m and n.

(2) Same as in section 11.

(3) Same as in section 11.

(4) Let d' and r' be the divergence and rise respectively for the related 1-jugate lattice with (m, n) phyllotaxis in the case of rhombic packing, or $(m, n; m+n)$ phyllotaxis in the case of hexagonal packing. Let R' be the radius of the cylinder for this 1-jugate lattice in the Erickson normalization. Then

$$R = kR'. \tag{19}$$

The internode distance h is the same on the original k-jugate lattice and the related 1-jugate lattice. Let δ' be the diameter of the spheres in the Adler normalization of the related 1-jugate lattice.

(5) Use the procedures of section 10 or section 11 for the cases of hexagonal or rhombic packing respectively. In each equation used in these procedures, first replace d by d', r by r', R by R', and δ by δ'. Then proceed with step (6) or step (7).

(6) For hexagonal packing, calculate d', r', δ', R' and h. Then use equation (19) to calculate R. Use equations (1) and (2) to calculate d and r.

(7) For rhombic packing, calculate δ', d', r' and h. Then use equations (1) and (2) to calculate d and r.

(8) The equations relating the parameters of the k-jugate lattice and the parameters of the related 1-jugate lattice are assembled here for convenience:

$$d' = kd, \quad r' = kr, \quad \delta' = k\delta, \quad R = kR'.$$

13. Some Calculated Results

For a variety of examples of hexagonal packing, which is the only type for which Erickson (1973) reports calculations, his results obtained by Method 1 are compared with results obtained by the method developed in this paper (Table 2). Erickson's paper gives no value of r, but it is implied

CONTACT PRESSURE AND CLOSE PACKING 457

TABLE 2

Comparison of results

	Erickson	Adler
$(x, y; x+y) = (1, 3; 4)$		
$360d$	97·743120	96·923077
h	0·4693787	0·480384
R	1·2905240	1·147683
r	0·057887	0·066617
$(x, y; x+y) = (1, 4; 5)$		
$360d$	77·414817	77·142857
h	0·3719648	0·377964
R	1·5712212	1·458679
r	0·037678	0·041239
$(x, y; x+y) = (2, 3; 5)$		
$360d$	141·843980	142·105263
h	0·3960824	0·397360
R	1·4862538	1·387481
r	0·042414	0·045580
$(x, y; x+y) = (4, 6; 10) = 2(2, 3; 5)$		
$360d$	71·023387	71·052632
h	0·3970983	0·397360
R	2·8223445	2·774962
r	0·022393	0·022790
$(x, y; x+y) = (3, 7; 10)$		
$360d$	107·068080	107·088608
h	0·1944493	0·194871
R	2·8799837	2·829200
r	0·010746	0·010962
$(x, y; x+y) = (4, 7; 11)$		
$360d$	98·721924	98·709677
h	0·1794356	0·179605
R	3·1136676	3·069669
r	0·009172	0·009312
$(x, y; x+y) = (2, 9; 11)$		
$360d$	162·513870	162·524272
h	0·1701343	0·170664
R	3·2795451	3·230493
r	0·008257	0·008408
$(x, y; x+y) = (3, 9; 12) = 3(1, 3; 4)$		
$360d$	32·335872	32·307692
h	0·4794166	0·480385
R	3·4865583	3·443048
r	0·021884	0·22206

458   I. ADLER

by his values for h and R via the formula $r = h/2\pi R$. Note that the results obtained by the two methods agree very closely for the higher values of $x+y$.

REFERENCES

ADLER, I. (1974). *J. theor. Biol.* **45**, 1.
ADLER, I. (1977). *J. theor. Biol.* **65**, 29.
BRAVAIS, L. & A. (1837). *Annls Sci. nat.* (*Bot.*) (2) **7**, 42; **8**, 11.
ERICKSON, R. O. (1973). *Science, Washington, D.C.* **181**, 705.
LE VEQUE, W. J. (1956). *Topics in Number Theory.* Reading, Mass.: Addison-Wesley.
VAN ITERSON, G. (1907). *Mathematische und Mikroskopische-Anatomische Studien der Blattstellungen.* Jena: Fischer.

JOURNAL OF ALGEBRA **205**, 227–243 (1998)
ARTICLE NO. JA977272

The Role of Continued Fractions in Phyllotaxis

Irving Adler

RR 1, Box 532, North Bennington, Vermont 05257

Communicated by Walter Feit

Received June 9, 1997

1. INTRODUCTION

Phyllotaxis is the study of the arrangement of botanical units such as leaves, scales, and florets around a stem. From the very beginning of the scientific study of this subject (in 1830; see (12)) the numbers observed in phyllotaxis were seen to be connected to simple continued fractions. The reason for this connection, however, was not immediately understood. Because of this, one leading botanist (see Section 8) rejected it outright as meaningless playing with numbers, and one well-known mathematician (see Section 7) was trapped into an erroneous argument leading to a false conclusion. It was not until 1974 that the real connection between continued fractions and the numbers significant in phyllotaxis was finally clarified in a few rigorously established theorems about cylindrical point-lattices. But first it was necessary to clarify the concepts that arise in phyllotaxis. This is the subject of Section 9. It was also necessary to seek out the underlying geometric meaning of a simple continued fraction. This is taken up in Section 12. The rest of this paper traces the history of the connection between phyllotaxis and continued fractions from the time it was first observed to the time it was finally understood.

2. BASIC CONCEPTS

The first detailed studies of phyllotaxis, performed by Schimper (12) and Braun (4, 5), were restricted at first to the arrangement of leaves around a mature stem and were then extended by Braun to the patterns formed by the scales of a pine cone. These studies introduced several basic concepts:

DEFINITION.   In the case where there is at most one leaf at any level on the stem the leaves may be pictured as points at equal intervals on a helix wound around a cylinder (Fig. 1). This helix is called the *fundamental* or *genetic spiral*.

DEFINITION.   A series of consecutive leaves forms a *cycle* if the highest leaf in the series is the first one to be directly over the lowest leaf.

Schimper and Braun assumed to begin with that such a cycle always exists.

DEFINITION.   The fraction of a turn between consecutive leaves in a cycle is called the *divergence*, and is designated by d.

When d is rational, as Schimper and Braun assumed it would be, it is equal to the fraction p/q, where p is the number of times the helix winds around the cylinder between the top and bottom of the cycle, and q is the number of intervals between consecutive leaves in the cycle.

Schimper and Braun took note of the fact that there are two different fundamental spirals that can be drawn for the same set of leaves, one joining them the short way around the stem and one joining them the

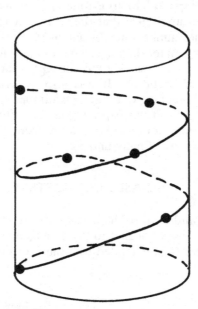

FIG. 1.   A cycle with five intervals in two turns: divergence $= 2/5$.

other way around. If the short way around is chosen, then the divergence is $\leq \frac{1}{2}$. The divergence, Braun asserted, determines all other properties of a leaf arrangement. This assertion turns out to be incorrect. Another number, the *rise*, designated by r and defined in Section 9, is also relevant.

3. THE FIBONACCI NUMBERS, THE GOLDEN SECTION, AND CONTINUED FRACTIONS ENTER THE PICTURE

In their observations of many plants, Schimper and Braun found that in the most common divergences p/q, p and q were terms of the Fibonacci sequence $\{F_n\}$ defined by the recurrence relation $F_{n+2} = F_n + F_{n+1}$ and the initial condition $F_1 = F_2 = 1$. For the short way around, the typical divergence has the form F_n/F_{n+2}. For the long way around, the divergence is F_{n+1}/F_{n+2}. Braun (5) observed that these fractions are related to certain continued fractions. The divergences measured the long way around are the convergents of the continued fraction

$$\cfrac{1}{1 + \cfrac{1}{1 + \cdots}}$$

and the divergences measured the short way around are the convergents of the continued fraction

$$\cfrac{1}{2 + \cfrac{1}{1 + \cdots}}.$$

In both fractions all terms after the first are equal to 1. The latter fraction converges to $1/(2 + \tau^{-1}) = \tau^{-2}$, where τ is the golden section, $(1 + \sqrt{5})/2$. They also noted that some divergences that occur less frequently are convergents of the continued fraction in which 2 is replaced by $t > 2$. Such a fraction converges to $1/(t + \tau^{-1})$. Thus, with these early observations, continued fractions entered into the study of phyllotaxis. To explain their intrusion into the subject, Braun said that nature preferred these particular continued fractions as a source of divergences because they were the simplest, in that all terms after the first were equal to 1.

His intuition that continued fractions were relevant turned out to be correct, as we shall see. The fact that all terms after the first are equal to 1 did, indeed, turn out to be significant, but his explanation did not really explain anything.

4. CONFUSION IN TERMINOLOGY

By assuming the existence of cycles, where the top leaf is directly above the lowest leaf, Schimper and Braun assumed that a divergence is always a rational number, that is, a number that can be expressed as the ratio of two whole numbers. But they also knew that this assumption was not always correct, so that the rational numbers they used as divergences could only be approximations. Moreover, they introduced some confusion by their use of the terms *rational* and *irrational*. Braun used the term rational to refer to unit fractions such as $\frac{1}{2}$ and $\frac{1}{3}$, and used the term irrational for all other fractions. However, in standard mathematical usage, *all fractions* are rational numbers, and irrational numbers are those that *cannot* be expressed as fractions. Because of this confusion of terminology he may have thought that he was taking irrational divergences into account.

5. BRAUN'S METHOD FOR CALCULATING d

The fundamental spiral is not always easy to see. In that case Braun said that the divergence can be calculated by using the orthostichies and parastichies that can be seen.

DEFINITION. *Orthostichies* are vertical alignments of the leaves.

DEFINITION. *Parastichies* are secondary spirals determined by joining leaves to other leaves that are not necessarily their neighbors on the genetic spiral.

The denominator q of the divergence p/q is the number of orthostichies. To obtain the numerator, p, he noted first that parastichies that cross each other form parallelograms. Starting with parastichies that have the least inclination to the horizontal, he drew the diagonals of the parallelograms that they are part of. These diagonals are part of a set of steeper parastichies. He next used these to form parallelograms whose diagonals yield still steeper parastichies. Continuing in this way, a chain of diagonals is formed the last of which is vertical. The number of diagonals in this chain yields the desired numerator p. Obviously this method of calculating the divergence works only if orthostichies are present, and they are present only if the divergence is rational.

This method is therefore not completely general. The general rule that connects the parastichies to the divergence was first discovered in 1974 and is described in Section 10. There are usually two sets of parallel parastichies that catch the eye. One set goes up to the left, and the other goes up to the right. The numbers of parastichies in these two sets are usually consecutive terms of the Fibonacci sequence. This is a second way in which the Fibonacci sequence enters into the study of phyllotaxis.

6. ADVANCES BY THE BRAVAIS BROTHERS

The Bravais brothers (6) were familiar with the work of Schimper and Braun, and referred to them frequently in their paper either by name or as "the German botanists." They, too, connected the most frequently observed divergences to the Fibonacci sequence and the continued fraction expansion for τ^{-2}. They argued, however, that the rational numbers cited by Schimper and Braun were not separate, distinct divergences, but only approximations to the one real divergence represented by the nonterminating continued fraction, namely the number τ^{-2}. Since then, botanists and mathematicians studying phyllotaxis have interpreted this in two different ways. Some assume that in the initial placement of leaves on the genetic spiral, they are already separated from each other by a divergence equal to τ^{-2}. Others, including this author, assume that other initial divergences are possible, but that there is a process occurring as the plant grows that causes the divergence to converge toward τ^{-2} as a limit.

The Bravais brothers used some techniques that have proved to be of lasting value in the further study of this subject. Picturing the plant stem as a cylinder and the leaves as points located at equal intervals on a helix, as did Schimper and Braun, they then introduced the plane development of the cylinder. This converts the genetic spiral and all parastichies into straight lines, and thus simplifies considerably the mathematics needed to describe the properties of the point-lattice. They numbered the leaves on the genetic spiral in the order of their appearance, making it possible to express phyllotactic relationships numerically. They also proved that if "the most easily seen" parastichies consist of m parallel spirals in one direction and n spirals in the opposite direction, then there is a single genetic spiral if and only if m and n are relatively prime. A refinement of the idea of what is "most easily seen" turns out to be necessary, and is taken up in Section 9.

7. A FALLACIOUS ARGUMENT BY TAIT

Tait (13), having seen the Bravais paper, decided that its elaborate arguments and calculations were superfluous. He offered what he considered to be a simple and complete solution to the puzzle of why the Fibonacci sequence seems to play a special role in phyllotaxis. If m parastichies are seen to go up to the left, and n parastichies go up to the right, with $m > n$ and m and n relatively prime, then there may also be seen a set of $m - n$ less steep parastichies going up to the left, also crossed by the n parastichies going up to the right. Continuing the process of subtracting the smaller number from the larger, one ultimately arrives at a single parastichy in one direction with some number t parastichies in the other direction. (This follows from the fact that m and n are relatively prime.) The single parastichy is the genetic spiral, and a single turn of it contains t but not $t + 1$ leaves. In the most common case, $t = 2$, and the divergence is necessarily $> \frac{1}{3}$ and $\leq \frac{1}{2}$. This will occur, for example, if you start with two consecutive Fibonacci numbers, say 8 and 13. The pairs obtained by the subtraction process that he proposes would be 8 and 5; then 3 and 5; then 3 and 2; and finally 1 and 2. He then imagines his subtraction procedure reversed, and concludes that whenever the divergence is between $\frac{1}{3}$ and $\frac{1}{2}$ the values of m and n for the most conspicuous spirals must be consecutive terms of the Fibonacci sequence. The fallacy of his argument is exposed in Section 10. Thompson (14), assuming that Tait's argument was valid, concluded that "the determination of the precise angle of divergence of two consecutive leaves of the generating spiral does not enter into the above general investigation ...; and the very fact that it does not so enter shows it to be essentially unimportant." This conclusion is the direct opposite of the statement by Braun cited at the end of Section 2. In Section 10, where we will state precisely how parastichy numbers are related to the divergence, we shall see that Thompson's conclusion is wrong.

8. SACHS REJECTS CONTINUED FRACTIONS AS NOT RELEVANT

Sachs (11), in his *Text Book of Botany*, pointed out that the continued fraction for $1/(2 + \tau^{-1})$ does not suffice to represent all divergences that are found in plants. Some are represented by the continued fraction for $1/(t + \tau^{-1})$, with $t > 2$. For this reason he concluded that "it seems to me absolutely impossible to imagine what value the method can have for a deeper insight into the laws of phyllotaxis." His reasoning here is faulty. He was ignoring the experience of physicists that in determining the laws governing a phenomenon two things are needed: a general rule, expressed in physics as a differential equation, and boundary conditions. In phyllotaxis the continued fraction for $1/(t + \tau^{-1})$ might well express a general rule, while the different values of t are the result of different boundary conditions. (See Sections 10–12.)

9. SOME NECESSARY DISTINCTIONS

The first step toward determining the exact role of continued fractions in phyllotaxis is discovering the precise connection between parastichy numbers and the divergence. But before this can be done it is necessary to introduce some distinctions among parastichies. For this purpose we begin with a cylindrical point-lattice with a single fundamental spiral going up to the right, numbering the lattice-points on this spiral $0, 1, 2, 3, \ldots$. Experience shows that the phenomena of phyllotaxis are independent of scale. To eliminate scale as a factor, we normalize the cylinder by taking the girth of the cylinder as unit of length. Let $0L$ be the element of the cylinder through leaf 0. Unroll the cylinder on a plane. Then the entire cylindrical lattice lies in a strip between two parallel lines, $0L$ and a copy of it, $0_1 L_1$. By repeating this strip over and over again to the left and right and extending the genetic spiral downward, we convert the cylindrical point-lattice into a point-lattice in the plane (Fig. 2). Note that it contains many copies of 0, designated respectively as 0_1, 0_2, 0_3, etc. In this picture, the divergence, d, of the genetic spiral is the horizontal component of the distance between two consecutive leaves on it.

DEFINITION.   The vertical component of the distance between two consecutive leaves on the genetic spiral on a normalized cylinder is called the *rise* and is designated by r.

(For a leaf distribution on a cylinder that is not normalized, the rise is the ratio of the internode distance to the girth of the cylinder.)

Let n be a leaf to the right of $0L$ whose distance from $0L$ is $\leq \frac{1}{2}$ and for which there is no leaf between 0 and n on the line that joins them. This line is a *right* parastichy; that is, it goes up to the right. It contains all those leaves and only those whose leaf numbers are multiples of n. Parallel to it are other parastichies, each containing the leaves belonging to a residue class modulo n. Thus the leaf n determines a set of n right parastichies. Similarly, a leaf m to the left of $0_1 L_1$ whose distance from $0_1 L_1$ is $\leq \frac{1}{2}$ and for which there is no leaf between 0_1 and m on the line that joins them determines a set of m left parastichies.

DEFINITION.   The m left parastichies and n right parastichies constitute an *opposed parastichy pair* and are designated by the ordered pair (m, n).

The m left parastichies cross the n right parastichies but, in general, there need not be a leaf at each of the intersections.

DEFINITION.   In the special case where there is a leaf at every intersection of an opposed parastichy pair, we call it a *visible opposed parastichy pair*.

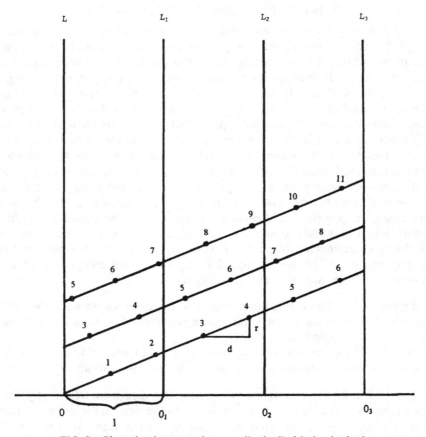

FIG. 2.   Plane development of a normalized cylindrical point-lattice.

These are the pairs that are relevant to our investigation. That they play a special role had already been recognized by Van Iterson (15) who called them "konjugierte spiralen."

We have one more definition to introduce:

DEFINITION.   If m and n are the leaves nearest to leaf 0 on the left and right, respectively, we call the opposed parastichy pair (m, n) *conspicuous.*

It is easily proved that a conspicuous opposed parastichy pair is a visible opposed parastichy pair. When botanists say that a leaf distribution has (m, n) phyllotaxis, they mean that (m, n) is a conspicuous opposed parastichy pair.

10. THE OPPOSED PARASTICHY TRIANGLE

We associate with any given opposed parastichy pair (m, n) a triangle constructed as follows: Extend the right parastichy determined by 0 and n to the lattice point mn. Then, from mn, draw downward the left parastichy determined by $0m$ that passes through it. This left parastichy will pass through some image 0_i of 0.

DEFINITION. We call the triangle whose vertices are 0, 0_i, and mn the opposed parastichy triangle belonging to (m, n).

The length of its base, 00_i is i. The following proposition is easily proved:

PROPOSITION 1. *An opposed parastichy pair is a visible opposed parastichy pair if and only if the base of its opposed parastichy triangle has length 1.*

We now introduce the concept of *contraction* that was used by Tait without defining it precisely.

DEFINITION. If (m, n) is a visible opposed parastichy pair with $m > n$, then $(m - n, n)$ is its *contraction*. If $n > m$, then $(m, n - m)$ is its contraction.

Then, using Proposition 1 and the elementary properties of a plane point-lattice, we can prove:

PROPOSITION 2. *The contraction of a visible opposed parastichy pair is a visible opposed parastichy pair.*

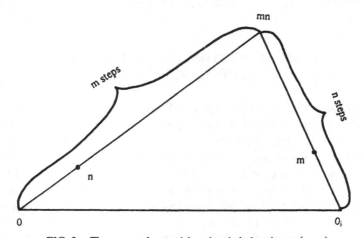

FIG. 3. The opposed parastichy triangle belonging to (m, n).

Tait sought to use the reverse of a contraction without realizing that a contraction can be reversed in two different ways, as shown in the following definition.

DEFINITION.   If (m, n) is a visible opposed parastichy pair, then $(m + n, n)$ is called its *left extension*, and $(m, m + n)$ is called its *right extension*.

The following three propositions proved in Adler (1), taken together, constitute what has been called the *Fundamental Theorem of Phyllotaxis* (8).

PROPOSITION 3.   *If (m, n) is a visible opposed parastichy pair, then there is a unique integer t such that (m, n) is the end result of a finite sequence of extensions starting with the visible opposed parastichy pair $(t, t + 1)$.*

PROPOSITION 4.   *$(t, t + 1)$ is a visible opposed parastichy pair if and only if $1/(t + 1) \le d \le 1/t$, where d is the divergence.*

PROPOSITION 5.   *Suppose that $[x/y, u/v]$ is the range of all possible values of d for which the opposed parastichy pair (v, y) is visible. Assume that x/y and u/v are in lowest terms. Let m be the mediant between x/y and u/v, namely $m = (x + u)/(y + v)$. Then the left extension of (v, y) is a visible opposed parastichy pair if and only if d is in the segment $[x/y, m]$, and the right extension of (v, y) is a visible opposed parastichy pair if and only if d is in the segment $[m, u/v]$.*

11. AN ALGORITHM FOR CALCULATING
THE RANGE OF d

Propositions 3, 4, and 5 provide an algorithm for determining the range of possible values of the divergence for any given visible opposed parastichy pair (m, n): Starting with (m, n) form successive contractions until you reach one of the form $(t, t + 1)$. Write these now in reverse order to obtain a sequence of extensions starting from $(t, t + 1)$. Write next to each extension L or R, to indicate whether it is a left or right extension of the pair that precedes it. Then use Proposition 5 to obtain the corresponding range of values of d. In the example below we use the algorithm to determine the range of possible values of d for the visible opposed parastichy pair $(34, 21)$:

Contractions	Extensions	Range of d
$(34, 21)$	$(2, 3)$	$[1/3, 1/2]$
$(13, 21)$	L $(5, 3)$	$[1/3, 2/5]$
$(13, 8)$	R $(5, 8)$	$[3/8, 2/5]$
$(5, 8)$	L $(13, 8)$	$[3/8, 5/13]$
$(5, 3)$	R $(13, 21)$	$[8/21, 5/13]$
$(2, 3)$	L $(34, 21)$	$[8/21, 13/34]$

Note that, contrary to Tait's assertion, each extension narrows the range of possible values of d.

12. THE GEOMETRIC MEANING OF A SIMPLE CONTINUED FRACTION

In Section 11 we saw how to determine the range of possible values of d for any given opposed parastichy pair that is visible. In order to proceed in the opposite direction, that is, to find which opposed parastichy pairs are visible for any given d, it is first necessary to understand the geometric meaning of the continued fraction for d. It is customary to introduce simple continued fractions either via the Euclidean algorithm or via Farey sequences. The geometric meaning of a simple continued fraction is implicit in the Farey sequence approach. It was made explicit for the first time by Adler (2), where it was shown that a simple continued fraction represents a medial nest of intervals.

To construct the medial nest that defines a particular number n, we proceed as follows: On the positive half of the real line designate 0 as $0/1$, and designate infinity as $1/0$. Insert the mediant between these two, namely $1/1$. It divides the half-line into two segments, namely $[0/1, 1/1]$ on the left, and $[1/1, 1/0]$ on the right. If n is in the left segment, write 0 as the first bit in a sequence of bits that will represent the nest of intervals that we are constructing to represent n. If n is in the right segment, write 1 as the first bit. Now, in the segment that contains n insert the mediant between its ends, thus dividing it into two segments, and write down as the second bit 0 or 1, according as n is in the left or right segment. Continue in this way, each time inserting the mediant between the ends of the segment that contains n, and writing 0 or 1 as the next bit in the sequence, according as n is in the left piece or the right piece of the segment just divided.

It is easily seen that the nested set of smaller and smaller intervals containing n is a genuine nest of intervals, and that n is the only number in the nest. The sequence of bits we have used to represent the nest therefore also represents the number n. (If n is irrational, there will be only one such non-terminating sequence. If n is rational, there will be two. One of the two contains only a finite number of 1's and continues after the last 1 with all 0's. The other contains only a finite number of 0's and continues after the last 0 with all 1's.) The sequence of bits may be seen as a sequence of clusters of 0's and 1's, as shown in the example below:

$$\underbrace{11111}_{a_0}\ \underbrace{00}_{a_1}\ \ \underbrace{111}_{a_2}\ \ \underbrace{0000}_{a_3}\ \underbrace{1111}_{a_4}\ \ \underbrace{00}_{a_5}\ \ \ldots$$

Let a_0 be the number of 1's in the first cluster. $a_0 \geq 0$. Let a_1 be the number of 0's in the second cluster, a_2 the number of 1's in the third cluster, etc. In general, a_i stands for a number of 1's if i is even, and for a number of 0's if i is odd. For $i > 0$ $a_i \geq 1$. Then $a_0 + 1/a_1 + 1/a_2 + 1/a_3 + 1/a_4 + \ldots$, (where everything that follows a division sign is understood to be under it), is the simple continued fraction for n. If the sequence a_i is terminating, its last term will be infinity. Leaving out the last term will give the continued fraction as it is usually written. However, the infinity should be retained for our purposes.

13. ONE ROLE OF THE CONTINUED FRACTION FOR d

Now let us assume that $n = d$, the divergence of a leaf distribution, which by definition is $\leq \frac{1}{2}$. Then $a_0 = 0$. If $a_1 = t \geq 1$, the first 0 in that cluster tells us that $d \leq 1$, the second 0 tells us that $d \leq \frac{1}{2}$, the third one tells us that $d \leq \frac{1}{3}, \ldots$, and the last 0 in the cluster tells us that $d \leq 1/t$. The first 1 that a_2 stands for tells us that $d \geq 1/(t + 1)$, so that $1/(t + 1) \leq d \leq 1/t$. Then, from Proposition 4 in Section 10, the opposed parastichy pair $(t, t + 1)$ is visible. By Proposition 2, all of its contractions, $(t, 1), (t - 1, 1), (t - 2, 1) \ldots (1, 1)$ are also visible. Now by Proposition 5, by starting with $(t, t + 1)$, we get further visible opposed parastichy pairs by taking first $a_2 - 1$ consecutive right extensions, then a_3 left extensions, then a_4 right extensions, etc., each set of a_i extensions for $i > 2$ being left or right according as i is odd or even. Thus we see that *the continued fraction expansion for the divergence d determines which opposed parastichy pairs are visible*. What we have outlined here is not restricted to divergences that actually occur in plants. It is a theorem of pure mathematics that applies to any cylindrical point-lattice generated by points placed at equal intervals on a single genetic spiral.

From the general algorithm developed here, others applicable to special cases have been derived (9).

14. TWO EXAMPLES

To illustrate what was developed in Section 13, we give two examples, one of a divergence that does not occur on any plants, and one that does occur on some. The divergence $d = \sqrt{3}/4$ does not occur on any plant, but can be used nevertheless on a single genetic spiral to generate a point-lattice on a cylinder. The simple continued fraction for this number is

$$0 + 1/2 + \overline{1/3 + 1/4}$$

where the bar indicates that the pair of terms $1/3 + 1/4$ is repeated over and over again ad infinitum. Then, according to Section 13, the visible opposed parastichy pairs associated with this value of the divergence are $(2, 3)$, and the consecutive contractions $(2, 1)$ and $(1, 1)$; then, starting from $(2, 3)$, two right extensions, namely, $(2, 5)$ and $(2, 7)$; then four left extensions, namely, $(9, 7)$, $(16, 7)$, $(23, 7)$ and $(30, 7)$; then three right extensions $(30, 37)$, $(30, 67)$ and $(30, 97)$; then four left extensions $(127, 97)$, $(224, 97)$, $(321, 97)$ and $(418, 97)$; etc., with three right extensions from here on alternating with four left extensions.

The divergence $d = 1/(3 + \tau^{-1})$ does occur on some plants. The simple continued fraction for this number is

$$0 + 1/3 + 1/1 + \overline{1/1}.$$

The visible opposed parastichy pairs associated with this value of the divergence are $(3, 4)$ and the consecutive contractions $(3, 1)$, $(2, 1)$, and $(1, 1)$; then, starting from $(3, 4)$, the extensions that are visible are alternately left and right, namely, $(7, 4)$, $(7, 11)$, $(18, 11)$, $(18, 29)$, etc.

15. A SECOND ROLE OF THE CONTINUED FRACTION FOR d

The opposed parastichy pair on a plant that catches the eye is the conspicuous opposed parastichy pair, defined in Section 9. It is determined by the two leaves that are nearest to leaf 0, one on the right and one on the left. When we try to identify which leaves are capable of qualifying as the leaves nearest leaf 0, we find another way in which the continued fraction for the divergence enters into the picture.

DEFINITION. There is a sequence of points $n_1 = 1, n_2, n_3, \ldots n_i, \ldots$ with the property that each n_i with $i > 1$ is the first lattice point with $n_i > n_{i-1}$ that is closer to the line $0L$ than n_{i-1} (Fig. 4). These were called "principal neighbors" by Coxeter (7) and "points of close return" by Adler (1).

The vertical component of the distance between leaf n_i and leaf 0 is m_i, where r is the rise. Since the horizontal component is smaller for this leaf than for any point of close return that precedes it in the sequence, it may become the leaf nearest leaf 0 if r is small enough. Thus d alone does not determine all the properties of a leaf arrangement, contrary to the statement by Braun cited in Section 2. The value of the rise r determines which of the n_i are the two leaves nearest leaf 0, and hence determines the

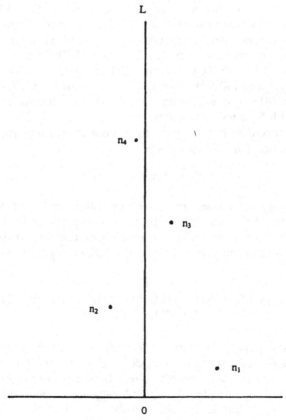

FIG. 4.   Points of close return.

conspicuous opposed parastichy pair. Coxeter (7) showed that the n_i are the denominators of the successive principal convergents of the expansion of d as a simple continued fraction.

16. A CAUSAL EXPLANATION

Nearly all observed divergences for a leaf distribution on a single genetic spiral have the form $1/(t + \tau^{-1})$. The continued fraction expansions for all these divergences have the common property that all their terms after the first are equal to 1. The Adler model of phyllotaxis provides a causal explanation of this fact. In the Adler model it is assumed that there is a period in the growth of a stem in which the rise is decreasing and the minimum distance between leaves is maximized. Under these conditions, it can be shown that the two leaves nearest leaf 0 must be equidistant from it, so that they lie on a circle with 0 as center. As r decreases, the next higher point of close return will join them so that there will be three points of close return on a circle with 0 as center (Fig. 5), and then with a further decrease of r it will displace one of them as leaf nearest 0. The three consecutive points of close return that are on the circle in Fig. 5 are the denominators of consecutive principal convergents of d, say, q_{n-1}, q_n, and q_{n+1}, and so are connected by the recurrence relation $q_{n+1} = q_{n-1} + a_{n+1}q_n$, where a_{n+1} is the term of the continued fraction for d that corresponds to the principal convergent p_{n+1}/q_{n+1}. If $a_{n+1} > 1$ there are leaves with leaf numbers greater than q_n that are intermediate neighbors on the segment that joins q_{n-1} and q_{n+1} (7), and that are not closer to $0L$ than q_n. But this is impossible if q_{n-1}, q_n and q_{n+1} are all equidistant from leaf 0. Hence under the assumption of maximization of the minimum distance between leaves, it is necessary that $a_{n+1} = 1$. As r continues to decrease, one term after another of the continued fraction for d is compelled to be equal to 1. While d is under this compulsion it alternately increases and decreases. The details of this process (1, 2) are not relevant to the purpose of this article and so are not given here.

17. PROPOSED FUNCTIONAL EXPLANATIONS

Why does nature have a preference for divergences in which the terms after the first in the continued fraction for d are all equal to 1? It has been proposed that the process that imposes this property on the values of d has been perfected by natural selection because it has survival value for the plant. However, there has been no consensus among botanists on what trait produced by this process has the postulated survival value. Two

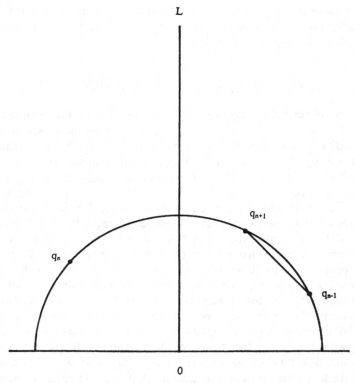

FIG. 5.   Three consecutive points of close return that are equidistant from 0.

different theories have been proposed invoking different properties alleged to follow from these values of d. One theory is that these values of d distribute the leaves around a mature stem in such a way as to minimize the shading of lower leaves by those above them, thus maximizing the amount of light they receive. Wiesner (16) claimed to have proved this theory experimentally. However, his experiment merely proved what was known in the first place, that upper leaves partially shade the lower leaves. His data did not show that the so-called "golden angles" shade them the least. Leigh (10) undertook a theoretical proof that this was so, but his argument was not conclusive in that it failed to take into account the influence of the internode distance. The other theory is that the advantage the golden angles give the plant is found at the growing tip of the stem, where the leaf embryos are crowded together, and not on the mature stem, where the leaves are widely separated and the internode distance is elongated. Airy (3) proposed this theory and demonstrated that close packing of equal spheres around a cylinder does indeed cause the numbers of conspicuous opposed parastichies to be consecutive Fibonacci numbers. In support of his theory he argued that "In the bud we see at once what must be the use of leaf order. It is for economy of space, whereby the bud is enabled to retire into itself and present the least surface to outward danger and vicissitudes of temperature."

18. SUMMARY

We have identified three ways in which the continued fraction for the divergence plays a role in phyllotaxis: (1) It determines which opposed parastichy pairs are visible. (2) The points of close return are the denominators of the principal convergents of the continued fraction. (3) When maximization of the minimum distance between leaves is in effect as r decreases, successive terms of the continued fraction are compelled to be equal to 1. It has been proposed in two different theories that this property of the divergence has survival value for a plant, but the arguments in support of these theories are not conclusive.

REFERENCES

1. I. Adler, *J. Theor. Biol.* **45** (1974), 1.
2. I. Adler, *J. Theor. Biol.* **65** (1977), 29.
3. H. Airy, *Proc. R. Soc.* **21** (1873), 176.
4. A. Braun, *Nova Acta Acad. Caesar. Leop. Carol.* **15** (1831), 197.
5. A. Braun, *Flora, Jena.* **18** (1835), 145.
6. L. Bravais and A. Bravais *Ann. Sci. Nat. Botan.* (2) 7, No. 42 (1837), 8, 11.
7. H. S. M. Coxeter, *J. Algebra* **20** (1972), 167.
8. R. V. Jean, "Mathematical Approach to Pattern and Form in Plant Growth," Wiley/Interscience, New York, 1984.
9. R. V. Jean, "Phyllotaxis. A Systematic Study in Plant Morphogenesis," Cambridge Univ. Press, Cambridge, 1994.
10. E. G. Leigh, The golden section and spiral leaf arrangement, *Trans. Conn. Acad. Arts Sci.* **44** (1972), 163.
11. J. Sachs, "Text-Book of Botany," Oxford Univ. Press, London, 1882.
12. K. F. Schimper, *Geiger's Mag. Fur Pharm.* **29** (1830), 1.
13. P. G. Tait, *Proc. R. Soc. Edinburg* **7** (1872), 391.
14. D. W. Thompson, "On Growth and Form," p. 930, Cambridge Univ. Press, Cambridge, 1968.
15. G. Van Iterson, "Mathematische und Mikroskopische-Anatomische Studien der Blatstellungen," Fischer, Jena, 1907.
16. J. Wiesner, *Flora, Jena.* **58** (1875), 113.

THE ROLE OF MATHEMATICS IN PHYLLOTAXIS

IRVING ADLER

North Bennington
Vermont 05257, USA

It is inevitable that mathematics enters into the study of symmetry in plants since "symmetry" is essentially a mathematical concept requiring mathematical methods to explore its properties. The perception of symmetry in plants has evolved with the passage of time from very general observations of the regular spacing of leaves (by Theophrastus, 370–285 B.C., and Pliny, 23–79 A.D.) to more and more specific and sophisticated descriptions of the properties of the spacing. That the spacing is often in a spiral arrangement was observed by Leonardo Da Vinci (1452–1519). That the Fibonacci numbers were somehow involved was a conjecture of Kepler (1571–1630). The first detailed study of the spiral arrangement of leaves was made by Schimper (1830), who introduced the concept of **divergence**, the angle between consecutive leaves on the spiral expressed as a fraction of a turn. For a cycle of leaves in which the last leaf is almost directly above the first one he defined the divergence as the number of turns around the stem divided by the number of intervals between leaves in the cycle. This definition automatically made every divergence a rational number, expressible as a fraction. Schimper also observed that the numerators and denominators of these fractions are terms of the Fibonacci sequence, thus confirming Kepler's hunch. The Bravais brothers (1837) argued that Schimper's rational divergences were only approximations of an underlying irrational divergence to which they converge. Thus the "ideal angle" $\tau^{-2}(360°) \approx 137.5°$, where τ is the golden section $(1 + \sqrt{5})/2$, entered into the picture. They also showed that if there are m conspicuous spirals going up to the left and n going up to the right, then m and n are relatively prime if and only if the leaves are on a single genetic spiral. If the greatest common divisor of m and n is $c > 1$ there are c genetic spirals and there are whorls of c leaves equally spaced around the stem at each node. The ground had finally been prepared for formulating the question that became the starting point of all later studies of phyllotaxis, namely, "Why in most plants with a single genetic spiral does the divergence converge to the ideal angle, and why are the terms m and n of the phyllotaxis (m, n) usually consecutive Fibonacci numbers?" Thus the Fibonacci sequence came to play a role in the study of phyllotaxis that is analogous to the role of the Balmer series (of lines in the spectrum of hydrogen) in the study of atomic structure.

Mathematics enters into the study of phyllotaxis in a variety of ways.

(1) **Organizing data.** This inevitably involves the use of statistical techniques. For example, Davies (1939) plotted a frequency distribution of 685 measured divergences in plants with normal phyllotaxis. By smoothing the curve, he obtained a curve resembling the normal curve of statistics. The mean divergence turned out to be 137° 39′ 57″, a very close approximation of the ideal angle. This fact clearly suggests that the ideal angle is involved in some way. But it does not tell us how it is involved, since several different interpretations of the statistics are possible. For example, (a) The divergence may be a constant, and the deviations from the value of the constant may be random errors of measurement; (b) The divergence may be the product of many independent factors each of which may or may not contribute to the measure of a particular case. This would be analogous to the frequency distribution for the number of heads that turn up when n coins are tossed; (c) The divergence may be a variable dependent on another parameter whose values have not been taken into account. In that case the frequency distribution, instead of helping to reveal the underlying cause, serves to hide it. These possibilities show that statistical techniques are not enough. They have to be supplemented by other methods of investigation.

(2) **Curve fitting.** When empirical data indicate that there is a functional relationship between two variables, a formula may be constructed that produces values that agree with the empirical data and also serves to interpolate additional values. This was done, for example, by Richards (1951) when he constructed a formula for the phyllotaxis index, that connects the level of phyllotaxis with the value of the plastochrone ratio. More generally, curve fitting is involved whenever a formula is sought to express an allometric relation.

(3) **Model building.** This is probably the most important way in which mathematics becomes involved in the subject. Constructing a model requires first making explicit the assumptions on which the model is based. The assumptions should, of course, have a biological justification. Then, secondly, the implications of these assumptions are deduced with appropriate rigor. In the last half century three different kinds of models of phyllotaxis have been constructed:

(a) *Descriptive models.* This type of model does not try to explain how or why phyllotaxis patterns acquire the form that they have, but merely examines the form to identify some distinguishing characteristics. This was done, for example, by

Coxeter (1972) when he pointed out that a divergence that produces a phyllotaxis pattern is distinguished by the fact that the continued fraction that represents it has no intermediate convergents. This observation became a significant clue in the development of Adler's contact pressure model (1974).

(b) *Cause and effect models.* In this type of model, some process is postulated, and then it is attempted to show that the results produced by the process correspond to observed phenomena. The classical example of this type of model is the one produced by Turing (1952) showing that the diffusion equation for the spread of a morphogen in a ring could produce a standing wave that might account for the formation of a whorl.

(c) *Evolutionary models.* In this type of model, some characteristic of phyllotaxis patterns is identified that is optimized in normal phyllotaxis. It is assumed that optimization of this property is of some advantage to a plant, and therefore accounts for its emergence in the course of evolution. An example of this type of model is the systemic model of Jean (1980).

One of the interesting facts that has come to light in recent years is that patterns similar to those of phyllotaxis arise in other fields of investigation quite remote from the growth of plants. Frey-Wyssling (1954) pointed out that a helical polypeptide chain displays characteristics analogous to those of a genetic spiral on a plant stem. Levitov (1991) found a phenomenon like phyllotaxis in a flux lattice in a layered superconductor. These developments suggest that plant growth, polypeptide chains, and superconducting layers have something in common that can account for the appearance of phyllotaxis patterns in all three domains. This is a question that requires further study.

In the Adler model (1974), the basic assumption used to explain the convergence of the divergence to the "ideal angle" is the maximization of the minimum distance between primordia. In the Jean model (1980) it is the minimization of the entropy. In the Levitov model (1991) and that of Douady and Couder (1992) it is the maximization of the energy of repulsion. The physicist Philip Morrison has suggested that it is likely that these three apparently different assumptions are mathematically equivalent. A proof of this conjecture would be a valuable contribution to the theory of phyllotaxis. It is hoped that someone reading this Prologue may be inspired to undertake the task to find such a proof.

There is one more comment that I think is appropriate at this time. No model of phyllotaxis, however well authenticated, can be the complete answer to the question of why plants display the symmetry patterns that they have. This is because a successful model not only supplies answers; it also raises questions. For example, Adler's contact pressure model shows that higher and higher normal phyllotaxis occurs as long as three conditions are fulfilled, (1) there is maximization of the minimum distance between primordia, (2) maximization begins early, and (3) the girth grows faster than the internode distance. But these conditions raise these questions for investigation: What causes the distance between neighboring primordia to tend toward maximization? What determines when maximization begins? What causes the variation in growth rates of the internode distance and the girth? And, more generally, how does all of this relate to what goes on in the tissues under the surface of the plant?

References

Adler I., A model of contact pressure in phyllotaxis, *J. Theor. Bio.* **45** (1974) 1–79.

Bravais L. and A., Essai sur la disposition des feuilles curviseriees, *Annals Sci. Nat. Bot.* **2** (1837) 42–110.

Coxeter H. S .M., The role of intermediate convergents in Tait's explanation for phyllotaxis, *J. Algebra* **20** (1972) 167–175.

Davies P. A., Leaf position in *Ailanthus altissima* in relation to the Fibonacci series, *Amer. J. Bot.* **26** (1939) 67–74.

Douady S. and Couder Y., Phyllotaxis as a physical self-organized growth process, *Phys. Rev. Lett.* **68** (1992) 2098–2101.

Frey-Wyssling A., Divergence in helical polypeptide chains and in phyllotaxis, *Nature* **173** (1954) 596.

Jean R. V., *Phyllotaxis* (Cambridge Univ. Press, 1994).

Levitov L. S., Phyllotaxis of flux lattices in layered superconductors, *Phys. Rev. Lett.* **66** (1991) 224–227.

Richards F. J., Phyllotaxis: Its quantitative expression and relation to growth in the apex, *Phil. Trans. R. Soc.* **B235** (1951) 509–561.

Schimper K. F., Beschreibung des Symphytum Zeyheri, *Geiger's Mag. Fur Pharm.* **29** (1830) 1–92.

Turing A. M., The chemical basis of morphogenesis, *Phil. Trans. R. Soc.* **B237** (1952) 37–72.

GENERATING PHYLLOTAXIS PATTERNS
ON A CYLINDRICAL POINT LATTICE

IRVING ADLER

North Bennington,
Vermont 05257, USA

1. The Problem

On the trunk of a palm tree, on a pine cone, on a pineapple, on the head of a sunflower, and, in general, at the growing tip of a stem, two sets of spirals can be observed, one set going up to the left and the other going up to the right. If there are m spirals going up to the left and n spirals going up to the right we say that the plant displays **(m, n) phyllotaxis**. These spirals that catch the eye are secondary spirals. The units (leaf primordia, scales, florets, etc.) that are arranged on these spirals lie on one or more primary spirals called **genetic spirals**. The number of genetic spirals is the number of units found at each level along the stem where one or more units are present. We shall assume that on each genetic spiral the units are arranged along it at equal intervals.

This chapter is concerned with identifying the features in the growth of a plant that can serve to explain three well-known facts:

(a) In the case where there is only one genetic spiral, m and n are nearly always consecutive terms of the Fibonacci sequences $\{1, 2, 3, 5, 8, ...\}$ which is generated by the rule that each term after the first two is the sum of the two that precede it.
(b) Also in this case the divergence, defined as the angle of rotation about the axis of the plant between two consecutive units, expressed as a fraction of a turn, rapidly converges to τ^{-2}, where $\tau = (1 + \sqrt{5})/2$ is the golden section.
(c) There are exceptions to the rules expressed in (a) and (b).

A model that serves to explain facts (a), (b) and (c) is developed in Secs. 10 to 21. But first, in Secs. 2 to 9, we construct the cylindrical representation of a system of phyllotaxis, and establish some of its general properties, culminating in what Jean (1984) called the Fundamental Theorem of Phyllotaxis.

This chapter is **not** concerned with determining when and where leaf primordia or other units first emerge. **This is a problem for botanists.** Our concern here is only with how their relative positions change after that, taking into account the interaction of several varying growth rates. On a cylindrical stem, four growth rates are relevant: the rate at which the number of primordia increases; the rate of growth of the primordia; the rate of growth of the girth of the stem; and the rate of growth of the internode distance, that is, the vertical distance between two consecutive units on the genetic spiral.

2. Choice of Surface

Phyllotaxis patterns occur on many different kinds of surface. To permit a uniform treatment of these patterns we shall use a cylindrical representation of each. First, each unit is represented by a point, which may be thought of as the center of the unit. Then the pattern of points is transferred from the original surface to the surface of a cylinder by means of an appropriate transformation (for disc to cylinder, see Adler, 1974; for cone to disc see Adler, 1977). With the help of these transformations a cylindrical representation can be obtained for the phyllotaxis pattern on any surface of revolution. For an arbitrary surface of revolution what we do in effect is divide the surface into narrow parallel zones between consecutive units. Each zone is approximately a zone of a cone. Then we transform cone to disc to cylinder (Fig. 1).

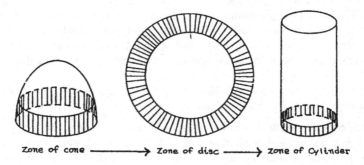

Zone of cone ⟶ Zone of disc ⟶ Zone of Cylinder

Fig. 1. Cylindrical representation of a zone on a surface of revolution.

The first investigators of phyllotaxis, e.g. Bravais and Bravais (1837), used a cylindrical representation because the surface they were principally interested in, the surface of a mature stem, was cylindrical. Under the influence of Church the cylindrical representation was abandoned in favor of the centric or disc representation in the interior of a disc. Church argued that phyllotaxis patterns were initiated at the growing tip of a stem, and that therefore cross-sections of the growing tip were the only appropriate surfaces to study. While he was right in looking for the origin of the patterns at the growing tip, he was wrong in rejecting the cylindrical representation. The cylindrical picture and the disc picture are mathematically equivalent via an appropriate transformation.

In this chapter we use the cylindrical representation because it simplifies considerably the mathematics used. The spirals seen in the disc picture are equiangular spirals that require transcendental equations to describe them. The corresponding spirals in the cylindrical picture are helices which, in the plane development of the cylinder, become straight lines requiring only simple algebraic equations.

3. The Normalized Cylinder

The phenomena of phyllotaxis occur on surfaces of different sizes, and apparently are independent of the size. To eliminate size as a factor we normalize the cylindrical representation by taking the girth of the cylinder as the unit of length. As unit of time T we take the **plastochrone**, the interval between the emergence of two consecutive units on the genetic spiral.

4. Development of the Cylinder

Let us designate by 0 any unit in a phyllotaxis pattern on a normalized cylinder. Imagine that we slit the cylinder along the line through 0 that is parallel to the axis of the cylinder and unroll the cylinder on a plane. The point 0 will appear in two places. Keep the designation 0 for the one on the left, and call the one on the right 0_1. The entire phyllotaxis pattern will be contained in a strip bounded by the vertical lines through 0 and 0_1 that represent the line along which the cylinder was cut. If this strip is repeated over and over indefinitely to the right and to the left, the points of the phyllotaxis pattern generate a point-lattice in the plane. There are more images of 0 to the right of 0_1. We continue the numbering by calling them 0_2, 0_3, etc.

5. Elementary Properties of the Point-Lattice

(1) If P and Q are lattice-points, then, if there are lattice-points between P and Q, they divide line segment PQ into congruent segments. Each of these segments is called a **step**.

(2) If P and Q are lattice points and there is no lattice-point between P and Q, and if R is a point on the line PQ whose distance from P is an integral multiple of the distance PQ, then there is a lattice-point at R.

(3) If P,Q,R and S are vertices of a parallelogram, and there are lattice-points at P,Q and R, then there is a lattice-point at S.

6. Parastichies

From now on we shall call each lattice-point a "leaf" to simplify the language and to remind us of the botanical context of our investigation. Let P be any leaf to the right of the vertical line through 0 such that there is no leaf between P and 0. The line determined by P and 0 is called a **right parastichy**. There are leaves on P0 at equal intervals equal to the distance P0. Through every leaf not on P0 draw a line parallel to it. We then have a set of parallel right parastichies. Similarly, if Q is a leaf to the left of the vertical line through 0_1, 0_1Q determines a set of parallel left parastichies. A set of m parallel left parasticies and a set of n parallel right parastichies is called an **opposed parastichy pair (m, n)**. In general, there need not be a leaf at each intersection of the lines of an opposed parastichy pair. If there is a leaf at every intersection in an opposed parastichy pair, the pair is called a **visible** opposed parastichy pair.

Proposition 1. If c is the greatest common divisor of m and n in a visible opposed parastichy pair (m, n), then there are c genetic spirals (Bravais, 1837).

Consequently, in the case where there is only one genetic spiral, m and n are relatively prime in every visible opposed parastichy pair (m, n).

From now on we restrict our attention to the case where there is only one genetic spiral. This entails no loss of generality, because the formulas for the case of n genetic spirals, $n > 1$, are easily derived from those for a single genetic spiral (Adler, 1974, p. 28). We shall consider only a genetic spiral that goes up to the right. Analogous results, obtained by symmetry, will apply to a genetic spiral that goes up to the left.

On the genetic spiral, the leaves are numbered with positive integers above 0 and negative integers below. The state of a system is determined by d and r, the horizontal and vertical components respectively of the geodesic distance between any two consecutive leaves i and i − 1. d is the divergence, the angle of rotation between i − 1 and i measured the shortest way around and expressed as a fraction of a turn. Hence $d \leq 1/2$. r is called the *rise*. If u is the girth of the stem and v is the internode distance, then r = v/u (Fig. 2). If 0 is chosen as the first leaf to appear then the leaves present at time T are 0, 1, ...,[T], where [T] stands for the greatest integer in T.

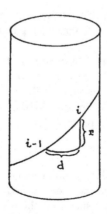

Fig. 2. Normalized cylindrical representation. Girth= 1.

If n is a leaf to the right of the vertical line through 0 such that there is no leaf between 0 and n, then n determines a set of n right parastichies, each containing the leaves whose numbers constitute a residue class modulo n. The parastichy that contains the leaf i, i = 0, 1, 2, ..., n−1, also contains i plus every multiple of n and no others. If m is a leaf to the left of the vertical line through 0 such that there is no leaf between 0 and m, a similar statement applies to the m left parastichies that it determines.

If (m, n) is a visible opposed parastichy pair, consider the parallelogram two of whose sides are 0m and 0n. The fourth vertex of this parallelogram is the leaf m + n (Fig. 3). There are no leaves in the interior of this parallelogram. In fact, if there were such an interior leaf P, there would be a parastichy through P parallel to 0n. It would intersect 0m at a point Q between 0 and m. Since (m, n) is a visible opposed parastichy pair, Q would be a leaf, contrary to our assumption that there is no leaf between 0 and m.

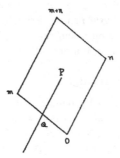

Fig. 3. Parallelogram.

7. Conspicuous Parastichies

A parastichy is called **conspicuous** if it is determined by joining a leaf to its nearest neighbor on the left or on the right. In an opposed parastichy pair, if both the left and the right parastichies are conspicuous, we call it a **conspicuous opposed parastichy pair.**

Proposition 2. A conspicuous opposed parastichy pair is a visible opposed parastichy pair. The proof is in Appendix A.

The spirals that catch the eye in a system of phyllotaxis represented by lattice-points are the conspicuous opposed parastichies. Hence, in the case where there is only one genetic spiral, if (m, n) is the phyllotaxis, m and n are relatively prime.

Which visible opposed parastichies can be conspicuous? To perceive the underlying relationships that provide an answer to this question, we first consider an artificial situation in which d is constant and r is decreasing. Later we shall see how the relationships are modified when, as a result of the dynamics of growth, both d and r are variable.

The conspicuous opposed parastichies are determined by the two leaves m and n that are nearest to leaf 0. We examine how their values are determined as r decreases. Let l_0 be the line through leaf 0 parallel to the axis of the cylinder. For n on the right, $dist(l_0, n)$, defined as the horizontal distance of n from 0, is equal to $nd - (nd)$, where (x) stands for the integer nearest to x. The vertical distance of n from 0 is nr. Then $dist^2 (0, n) = dist^2(l_0, n) + n^2r^2$. For m on the left we have $dist(l_0, m) = (md) - md$, and $dist^2(0, m) = dist^2(l_0, m) + m^2r^2$.

We define a sequence of **points of close return** as follows. Let n_1 be the leaf nearest leaf 0. For i = 1, 2, ... let n_{i+1} be the first leaf with higher leaf number than n_i that is closer to l_0 than n_i. As r decreases, the vertical component of $dist^2 (0,n_i)$ shrinks so that the value of $dist^2(0,n_i)$ is determined principally by the horizontal component. As a result, n_1 is displaced in its role as leaf nearest 0 by n_2, whose distance from leaf 0 has a smaller horizontal component. Then n_2 is displaced by n_3, etc., each of the points of close return taking its turn as leaf nearest 0 as r decreases. In the dynamic situation taken up in Sec. 15, where d is also variable, we shall be able to identify the leaf numbers of the n_i.

8. The Opposed Parastichy Triangle

Let (m, n) be an opposed parastichy pair, not necessarily visible. In the plane point-lattice described in Sec. 4 we associate with (m, n) a triangle constructed as follows. Draw the line through 0 and n. This is one of the right parastichies determined by n. On it lie the lattice points that are multiples of n. In particular the lattice point mn is on it. Since mn is a multiple of m, it also lies on a left parastichy determined by m. In fact it lies on the left parastichy which contains leaf m and leaf 0. Therefore when it is drawn downward from mn, it will meet one of the images of 0, 0_i. The triangle whose vertices are 0, 0_i and mn is called the **opposed parastichy triangle** of the opposed parastichy pair (m, n) (Fig. 4). (This definition differs slightly from that given by Adler, 1974.) The opposed parastichy triangle of the opposed parastichy pair (m, n) has the following properties: Its base

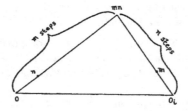

Fig. 4. Opposed parastichy triangle for the opposed parastichy pair (m, n).

has length i; Its vertex is mn; On the left leg, the first leaf above leaf 0 is n; The leaves on the left leg divide it into m steps; On the right leg the first leaf above 0_i is m; The leaves on the right leg divide it into n steps.

Proposition 3. The opposed parastichy pair (m, n) is visible if and only if the base of its opposed parastichy triangle has length 1. The proof is given in Appendix B. An opposed parastichy triangle whose base has length 1 is called a **visible opposed parastichy triangle**.

An algebraic version of Proposition 3 can be derived as follows: The projection on the base of the left leg of the opposed parastichy triangle has length m{nd − (nd)}. The projection on the base of the right leg has length n{(md) − md}. The sum of the two projections is the length of the base, which is 1 if and only if the opposed parastichy pair is visible. Therefore m{nd − (nd)} + n{(md) − md} = 1 if and only if the opposed parastichy pair is visible. Simplifying the left-hand side of this equation we get

Proposition 3′. The opposed parastichy pair (m, n) is visible if and only if n(md) − m(nd) = 1.

9. Contraction and Extension

In Sec. 1 we described the phyllotaxis and the divergence of a phyllotaxis system as if they were independent of each other. They are, in fact, closely related to each other. In this section we determine how they are related.

If (m, n) or (n, m) with m > n is a visible opposed parastichy pair, the opposed parastichy pair (m − n, n) or (n, m − n) respectively is called its **contraction**.

Proposition 4. The contraction of a visible opposed parastichy pair is a visible opposed parastichy pair (The proof is in Appendix C).

Proposition 5. If (m, n) is a visible opposed parastichy pair with m − n > 1 and n > 1, then there is a unique t > 1 such that (t, t + 1) is the visible opposed parastichy pair that is the end result of finitely many successive contractions of (m, n). This follows immediately from the fact that m and n are relatively prime.

Proposition 6. For every t > 1, (t, t + 1) is a visible opposed parastichy pair if and only if 1/(t + 1) ≤ d ≤ 1/t (The proof is in Appendix C).

The reverse of a contraction is called an **extension**. While a visible opposed parastichy pair (m, n) has only one contraction, it has two extensions. We call (m + n, n) its **left extension**, and (m, m + n) its **right extension**. The following proposition gives the conditions that determine whether one or the other of the extensions is a visible opposed parastichy pair.

Proposition 7. Suppose that [x/y, u/v] is the range of all possible values of d for which the opposed parastichy pair (v, y) is visible. Assume that x/y and u/v are in lowest terms. Let m be the mediant between x/y and u/v, namely m = (x +

u)/(y + v). Then the left extension of (v, y) is a visible opposed parastichy pair if and only if d is in the segment [x/y, m], and the right extension of (v, y) is a visible opposed parastichy pair if and only if d is in the segment [m, u/v]. (The proof is in Appendix C.)

Propositions 5, 6 and 7 together constitute the Fundamental Theorem of Phyllotaxis. They provide an algorithm for answering these two questions: 1. If (m, n) is a visible opposed parastichy pair, what is the range of all possible values of d? 2. For a given value of d, what are all possible visible opposed parastichy pairs?

Example. Find the range of possible values of d if (9, 7) is a visible opposed parastichy pair. The succesive contractions of (9, 7) are (2, 7), (2, 5), (2, 3). In the table below they are listed in reverse order, starting with (2, 3). L written to the left of a pair indicates that it is a left extension of the pair that precedes it. R indicates a right extension. The column on the right shows the corresponding range of values of d.

Table 1.

	Pair	Range of values of d
	(2, 3)	[1/3, 1/2]
R	(2, 5)	[2/5, 1/2]
R	(2, 7)	[3/7, 1/2]
L	(9, 7)	[3/7, 4/9]

Therefore (9, 7) is a visible opposed parastichy pair if and only if the divergence lies in the segment [3/7, 4/9].

Example. Find all visible opposed parastichy pairs if d = 3/8. First we determine the consecutive unit fractions between which 3/8 lies. They are 1/3 and 1/2. Therefore (2, 3) is a visible opposed parastichy pair. The mediant between 1/3 and 1/2 is 2/5. Since 3/8 lies to the left of 2/5, in the segment [1/3, 2/5], the left extension of (2, 3), namely (5, 3), is a visible opposed parastichy pair. The mediant between 1/3 and 2/5 is 3/8. Since 3/8 lies in both segments [1/3, 3/8] and [3/8, 2/5], both left and right extensions of (5, 3) are visible, namely (8, 3) and (5, 8). When we insert more mediants starting with the segment [1/3, 3/8], we get in succession [4/11,3/8], [7/19, 3/8], etc., with 3/8 always on the right. Therefore a succession of right extensions of (8, 3) are also visible, namely (8, 11), (8, 19), ... Similarly a succession of left extensions of (5, 8) are also visible, namely (13, 8), (21, 8), ...

For any t ≥ 2, the tree diagram in Fig. 5 shows the possible visible extensions of (t, t + 1) and the corresponding ranges of d.

In view of the fact that a simple continued fraction represents a mediant nest of intervals, the continued fraction for the divergence d is intrinsically involved in the identification of the visible opposed parastichy pairs determined by d. For details, see Appendix D in Adler, 1977.

10. Kinds of Explanation Sought

In trying to explain the facts of phyllotaxis, two different kinds of explanation may be sought. One is **causal**, identifying causes capable of producing the phenomena of phyllotaxis as their effects. The other is **functional**, showing how the phyllotaxis patterns are of use to the plant. Our choice in this chapter is to give a causal explanation, although, as we shall see in Sec. 20, a functional interpretation of this explanation is also possible.

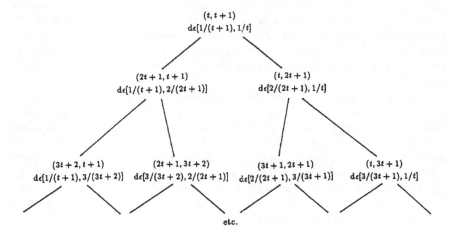

Fig. 5. Visible extensions of $(t, t+1)$ and the corresponding ranges of d.

11. How a Model is Constructed

A model of phyllotaxis should have the following characteristics:

1. Certain assumptions are made.
2. The assumptions should be based on evidence derived from botany.
3. The consequences of the assumptions are deduced.
4. It is shown that the phenomena for which an explanation is sought are among the consequences.
5. Further predictions made by the model suggest observations or experiments to be made to check the validity of the model.

12. The Assumptions

(a) As already stated in Sec. 1, we assume that the leaves are arranged along the genetic spiral at equal intervals.

(b) We assume that there is a period in the growth of the plant when the rise r is decreasing. Since the rise is the internode distance divided by the girth, this simply means that there is a period when the girth grows faster than the internode distance.

(c) We assume that beginning with some time T_c there is a period when the minimum distance between leaves is maximized.

Assumption (c) was originally formulated as a way of giving some precision to Schwendener's assumption of contact pressure (Schwendener, 1878 and Adler, 1974). A heuristic argument to justify it was given as follows: Any leaf and its nearest neighbor, as they grow in size, grow towards each other until they touch. As they continue to grow after making contact, their centers move apart. Consequently the minimum distance between leaves begins to increase. However, since the leaves are confined to a finite space, it cannot increase indefinitely, and hence attains a maximum value.

Some botanists have rejected the model based on these assumptions on the grounds that contact pressure does not exist in ferns. The rejection is based on a misunderstanding of the nature of a mathematical model. The model does not assert that the assumptions apply to all plants. What it does assert is that **where** the assumptions do apply, certain consequences follow. Since there are so many plants where the growing primordia do make contact, the relevance of the model cannot be questioned. Moreover, assumption (c) is not necessarily limited to situations where there is contact pressure. Any process that causes a repulsion of neighboring leaves is consistent with assumption (c).

13. Plotting Minimum Distance as a Function of d

Because of Propositions 5 and 6 in Sec. 9, there is a natural division of the interval $(0, 1/2]$ (open on the left and closed on the right) into segments between consecutive unit fractions, $[1/(t + 1), 1/t]$, $t \geq 2$. As a plant grows, the divergence d and the phyllotaxis undergo some changes. The precise nature of these changes depends on which of these segments d lies in initially. Since in most plants d lies in the segment $[1/3, 1/2]$, this is the interval we shall deal with in this section and those that follow. A complete generalization for $t > 2$ of what is developed in this section is given in Appendix D.

To get some clues to the implications of the assumptions, we look into how the minimum distance varies with d as r decreases and T increases, when d is between $1/3$ and $1/2$. Since the minimum distance and its square will attain a maximum

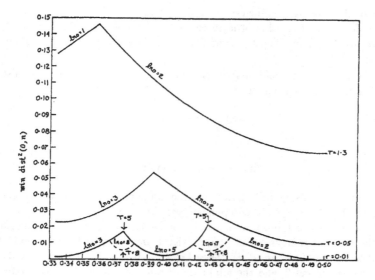

Fig. 6. Min dist^2(0 ,n) as a function of d when T = 5 and T =8 (lno=leaf nearest zero).

together, it suits our purpose to use the square of the minimum distance and thus avoid superfluous calculation. The plot is shown in Fig. 6 for three values of r, namely 1.3, 0.05, and 0.01. In the third case a separate curve is required for T = 5 and T = 8.

An examination of the graph reveals the following characteristics: 1. For high values of r either leaf 1 and leaf 2 are the nearest neighbors to leaf 0, or leaf 2 and leaf 3, and this fact is independent of the value of T as long as these leaves are present. 2. When r is low enough, the value of T becomes relevant. 3. When r is low enough, the minimum distance is maximized at the value of d for which the two nearest neighbors of leaf 0 are equidistant from it. 4. As T increases and r decreases there are more and more maxima. In the graph, for example, we see a transition from one maximum to two and then four. 5. When a later leaf displaces one of the two leaves nearest 0 in its role of leaf nearest 0, the new leaf number is the sum of the other two leaf numbers. In the sections that follow we give independent proofs of these statements, and explore their consequences.

14. If r Is Small Enough, and the Minimum Distance Between Leaves is Maximized, then the Leaves Nearest 0 are Equidistant from It

This statement, derived from observation (3) above, is proved in Appendix D. We shall specify below when it is that r is "small enough." We saw in Sec. 7 that the leaves capable of being "leaf nearest leaf 0" are those at the points of close return, n_i. As r decreases, the value of n_i increases. The lowest n_i's that can be leaves nearest 0 are 1 and 2. Let us now apply the condition that they are equidistant from leaf 0.

$\text{dist}^2(0, 1) = d^2 + r^2$, and $\text{dist}^2(0, 2) = (1 - 2d)^2 + 4r^2$. Setting these distance squares equal, we get the equation

$$(1 - 2d)^2 + 4r^2 = d^2 + r^2,$$

which yields

$$(d - 2/3)^2 + r^2 = (1/3)^2. \tag{1}$$

This equation tells us, first, that when the minimum distance between leaves is maximized, d and r cease to be independent variables. Instead d becomes a function of r. To see what this implies we make use of the (d, r) plane, the phase space in which the state of the system of phyllotaxis at any moment is represented by a point whose coordinates are (d, r). Then Eq. (1) tells us that if leaves 1 and 2 are the leaves nearest leaf 0, and the minimum distance between leaves is maximized, then, if r is small enough, the point (d, r) is confined to the circle whose radius is 1/3 and whose center is at (2/3, 0). Since r > 0, the point (d, r) is really confined to the semicircle above the d axis. We call it the **(2, 1) semicircle**. Since d < 1/2, it is really confined to the arc of the semicircle that lies to the left of the vertical line d = 1/2. As r decreases, the point (d, r) is compelled to descend along this arc, until another leaf with a higher leaf number becomes close enough to leaf 0 to displace leaf 1 as leaf nearest 0. We will determine the identity of this leaf in the next section.

It is clear from Fig. 7 that when leaves 1 and 2 are the leaves nearest 0 and are equidistant from 0, the highest possible value of r occurs when d = 1/2. Substituting 1/2 for d in Eq. (1) we find that this value of r is $\sqrt{(1/12)}$. "Low enough" in the statement above means not greater than $\sqrt{(1/12)}$. If $r > \sqrt{(1/12)}$, it is impossible for leaves 1 and 2 to be equidistant from leaf 0. Then only leaf 1 is nearest to leaf 0, and its distance from leaf 0 is maximized when d = 1/2. If maximization of the minimum distance between leaves occurs when $r > \sqrt{(1/12)}$, then (d, r) is on the line d = 1/2, and successive leaves on the genetic spiral are separated by 180° around the axis of the stem. As r decreases, the point (d, r) descends along the line d = 1/2 until it reaches the (2, 1) semicircle. Then as r decreases further, (d, r) descends to the left along the semicircle.

15. The Addition Rule

Figure 6 suggests that when, with decreasing r and maximization of the minimum distance between leaves, a later leaf displaces one of the two leaves nearest 0 in its role of leaf nearest 0, the new leaf number is the sum of the other two leaf numbers.

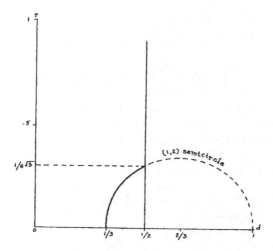

Fig. 7. Highest possible value of r.

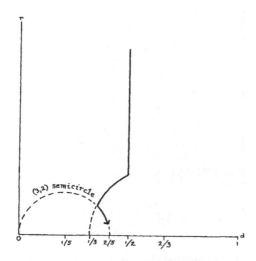

Fig. 8. Transition from the (2, 1) to the (2, 3) semicircle.

A proof of this rule is given in Appendix D. When the point (d, r) descends along the (2, 1) semicircle, it ultimately reaches a value of r at which leaf 3 will displace either leaf 1 or leaf 2 as leaf nearest 0. Figure 6 shows that leaf 1 is the leaf that is displaced, so that (2, 1) phyllotaxis is replaced by (2, 3) phyllotaxis, and the point (d, r) begins to descend on the (2, 3) semicircle defined by the condition that dist^2(0, 3) = dist^2(0, 2). This condition leads to the equation (d $-1/5)^2 + r^2 =$ $(1/5)^2$, so we see that the (2, 3) semicircle has center (1/5, 0) and radius 1/5. The transition from the (2, 1) to the (2, 3) semicircle is shown in Fig. 8.

As the point (d, r) descends along the (2, 3) semicircle it ultimately reaches a value of r at which leaf 5 displaces either leaf 3 or leaf 2 as leaf nearest 0. Which one will it displace? We find a clue to the answer in Fig. 6. Once leaf 5 is capable of being a leaf nearest leaf 0, the curve plotting the square of the minimum distance has at first two maxima. Maximization of the minimum distance makes each of these maxima an **attractor**. Figure 6 suggests that if d < 0.4, the effective attractor is the one where leaves 3 and 5 are the leaves nearest leaf 0, and if d > 0.4 the effective attractor is the one where leaves 2 and 5 are the leaves nearest leaf 0. But when the point (d, r) in the phase space is confined to the (2, 3) semicircle, we see in Fig. 8 that d < 0.4. Therefore leaf 5 displaces leaf 2, and the point (d, r) begins to descend along the (5, 3) semicircle, whose equation is (d $-7/16)^2 + r^2 = (1/16)^2$. This is confirmed by the fact that where the two semicircles intersect, d is approximately equal to 0.39474, which is less than 0.4 (see Fig. 9).

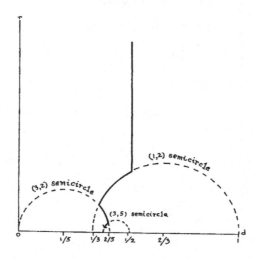

Fig. 9. Zig-zag path with each succeeding arc smaller than the one that precedes it.

If r continues to decrease, the point (d, r) will ultimately reach and move down the (5, 8) semicircle, and then the (13, 8) semicircle, etc., provided that each pair of consecutive semicircles intersects (The proof that they do indeed intersect is given in Appendix D). This is a zig-zag path with each succeeding are smaller than the one that precedes it. The projection of these arcs on the d axis is a nest of intervals that converges to d = 1/(τ + 1) = τ^{-2}, where τ is the golden section. The path just described is the **normal phyllotaxis path** producing the kinds of phyllotaxis that are found most often. However it is not the only phyllotaxis path that is possible.

16. More and More Maxima

In Fig. 6, were we to plot more curves for lower and lower values of r, we would see more and more points of close return coming into play as leaf nearest 0, and in consequence there would be more and more maxima on these curves as r decreases. If maximization of the minimum distance doesn't begin until these multiple maxima are present, the maximum which serves as an effective attractor is determined by what the value of d is at the time that maximization of the minimum distance goes into effect. If, for example, maximization of the minimum distance begins at time $T = 8$ when $r = 0.01$, any one of four maxima may be the effective attractor. If d is in an interval near and surrounding the value of d at which a maximum occurs, it will tend to move toward that value. Then each maximum can be the beginning of a zig-zag phyllotaxis path governed by the addition rule. On each such path the phyllotaxis rises to higher and higher values taken from consecutive terms of a Fibonacci-type sequence generated by the recurrence relation that each term is the sum of the two that precede it. However, the recurrence relation alone is not enough to guarantee that the sequence will be the original Fibonacci sequence. A Fibonacci-type sequence will be **the** Fibonacci sequence if and only if its first two terms are themselves Fibonacci numbers. For example, if maximization of the minimum distance begins at a time when $d = 0.42$ and the leaves nearest leaf 0 are 2 and 5, then the point (d, r) will first descend on the $(2, 5)$ semicircle, then the $(7, 5)$ semicircle, etc. The sequence associated with this phyllotaxis path is $\{2, 5, 7, 12, 19, ...\}$.

17. When Maximization Begins Early

Figure 6 indicates that more than one maximium occurs in the curve for the square of the minimum distance only when r is low enough and T is high enough. This is confirmed by a detailed analysis in Adler, 1977. If $T < 5$, there is only one maximum in the curve, and the only possible semicircles that are available for starting a phyllotaxis path are the $(2, 1)$ semicircle and the $(2, 3)$ semicircle. These are also the only ones available even if $T > 5$, if r is greater than the value at which the $(2, 3)$ semicircle intersects the $(5, 3)$ semicircle and the $(2, 5)$ semicircle. This value turns out to be $(\sqrt{3})/38$. Consequently, if maximization of the minimum distance between leaves begins early, where early is defined to mean before $T = 5$ or when $r > (\sqrt{3})/38$, then normal phyllotaxis is inevitable, since both $(1, 2)$ and $(2, 3)$ are consecutive Fibonacci numbers.

18. A Computer Program for the Model

A computer program for the cylindrical representation of a disc model of phyllotaxis is given in Appendix E. An appropriate formula for r as a function of T is given. The program is based only on the assumptions, so it provides independent confirmation of the conclusions we have drawn. For each value of T starting with $T = 2$ it calculates the distance from leaf 0 of each leaf that is present at that time, and then identifies the two leaves that are nearest to leaf zero. At $T = 2$, an initial value of d is assumed. Then d is gradually changed in the direction in which the distance from leaf 0 of the leaf nearest 0 will be increased until it is maximized. The value of d that maximizes the minimum distance at $T = n$, ($n = 0, 1, 2, 3, ...$), becomes the initial value of d for $T = n + 1$, and the same procedure is followed for each value of T. After each calculation it plots on a graph in the (d, r) plane the point representing the (d, r) attained, and also prints out for each value of T the leaf numbers of the two leaves nearest 0 and the value of d that maximizes the minimum distance between leaves. Table 1 in Appendix E is the printout from such a program. Table 2 is the printout for the cylindrical representation of a parabolic model, using a formula for r that is appropriate for a paraboloid. In Table 1, maximization of the minimum distance between leaves was assumed to begin at $T = 2$, with the initial value of d between 1/3 and 1/2. The table shows that the result is normal phyllotaxis. As T increases, higher and higher phyllotaxis is attained, but always using consecutive terms of the Fibonacci sequence, $\{1, 2, 3, 5, 8, ...\}$. In Table 2, maximization of the minimum distance between leaves was assumed to begin at $T = 5$, with the initial value of d between 2/5 and 1/2. In this case, as higher and higher phyllotaxis (m, n) is attained, m and n are consecutive terms of the sequence $\{2, 5, 7, 12, ...\}$.

19. Weakening the Assumptions

One of our assumptions was that the leaves emerge at equal distances on the genetic spiral. This assumption turns out to be superfluous when we assume maximization of the minimum distance between leaves. Let r_i be the vertical component of the distance between leaves $i - 1$ and i. Let d_i be the horizontal component. If the r_i are not all equal to some number r, the vertical component of the distance between leaves 0 and n is $r_1 + r_2 + ... + r_n$ and takes the place of nr in the calculation of the distance. Calculations show that the only effect this has is to alter the rate at which the advance to higher phyllotaxis takes place. On the other hand, maximization of the minimum distance between leaves compels the equalization of the d_i. For the proof of this assertion, see Appendix F in Adler, 1977, page 77.

20. A Functional Interpretation

The assumption of maximization of the minimum distance between leaves is equivalent to the assumption of close packing. In fact, if equal spheres are close-packed around a cylinder, the two nearest neighbors of any sphere are equidistant from it, which is also a consequence of the assumption of maximization of the minimum distance. Airy proposed the theory that close-packing of leaf primordia at the tip of a stem was advantageous to the plant because "the bud is enabled to retire into itself and present the least surface to outward danger and vicissitudes of temperature" (Airy, 1873).

21. Predictions Made by the Model

The model developed here makes several predictions. They offer a challenge to botanists to develop methods of observation by which the predictions may be either verified or refuted.

1. The dynamo that drives the advance of the phyllotaxis of a plant from lower phyllotaxis (small numbers of conspicuous spirals) to higher phyllotaxis (greater numbers of conspicuous spirals) is the decrease of r. Since r is the internode distance divided by the girth of the stem, this advance will take place only as long as the girth grows faster than the internode distance.

2. The advance to higher and higher phyllotaxis takes place as the point (d, r) representing the state of the system in the (d, r) plane descends along the zig-zag path we have called the phyllotaxis path. In the course of this descent the value of d keeps changing, alternately decreasing and increasing.

3. Each arc of this path is part of some (m, n) semicircle, where (m, n) is the phyllotaxis displayed while (d, r) is on that arc. If d and r can be measured, they should satisfy the equation for the (m, n) semicircle. For example, while the phyllotaxis is (3, 5) the d and r should satisfy the equation $(d - 7/16)^2 + r^2 = (1/16)^2$.

4. While (d, r) is on the arc of a particular (m, n) semicircle, the range of possible values of d is smaller than the range of values for which (m, n) can be a visible opposed parastichy pair. In fact d is restricted to the interval between the values of d at which the semicircle intersects the semicircles for the next lower and next higher phyllotaxis. For example, (5, 3) is visible as long as d is in the interval [1/3,2/5]. However, the (5, 3) semicircle intersects the (2, 3) semicircle at d = 15/38, and it intersects the (5, 8) semicircle at d = 23088/61152. So while (d, r) is on the (5, 3) semicircle d is restricted to the smaller range between these two limits. (The lower limit of the range is approximately .37758, and the upper limit is approximately .39474.)

5. Let D be the maximized minimum distance between leaves. Then it is also the relative leaf diameter, that is, the diameter expressed in terms of the girth of the stem as unit of length. On the (2, 1) semicircle, $D^2 = d^2 + r^2$, and $D^2 = (1-2d)^2 + 4r^2$. If we eliminate r from these two equations, we find that $D^2=(4/3)d-1/3$. By a similar procedure we find that on the (2,3) semicircle, $D^2 = 1 - (1/12)d$. In general, on each arc of the phyllotaxis path, D^2 is a linear function of d, but it is a different function of d on each arc. If we assume the leaves to be circles, the relative area of each would be $\pi D^2/4$. Hence the relative area of the leaves is a linear function of d on each arc of the phyllotaxis path. As r decreases, d decreases on the (1, 2) semicircle, and so does D^2. As r decreases, d increases on the (2, 3) semicircle, but because of the form of the corresponding equation for D^2, D^2 continues to decrease. In general while d alternately decreases and increases on the consecutive arcs of the phyllotaxis path, the relative area of the leaves (expressed in terms of the girth of the stem as unit of length) always decreases.

Acknowledgments

I am grateful to Peggy Adler for the Figures. Figure 6 is reproduced with permission from the *Journal of Theoretical Biology* **65** (1977). Copyright: Academic Press.

Appendix A

Proposition 2. A conspicuous opposed parastichy pair is a visible opposed parastichy pair.

Proof. In the conspicuous opposed parastichy pair, let P be the intersection of a left parastichy and a right parastichy, and assume that there is no leaf at P. We show that this assumption leads to a contradiction. On the left parastichy let A and C be the leaves nearest P on opposite sides of P. On the right parastichy, let B and D be the leaves nearest P on opposite sides of P (see Fig. A1). By hypothesis, (that the opposed parastichy pair is conspicuous), dist(A, C) and dist(B, D) are the two smallest distances between leaves. If they are not equal, we may assume without loss of generality that dist(A, C) is the larger of the two. Let y = dist(A, C). There must be two of the four leaves, say A and B, such that dist(A, P) = ry and dist(B, P) = sy, with r, s $\leq$ 1/2. Let Q be angle APB, and x = dist(A, B). Since dist(A, C) and dist(B, D) are the two smallest distances between leaves, and since AB is not parallel to BD or AC, then x $\geq$ y.

$$x^2 = r^2y^2 + s^2y^2 - 2rsy^2 \cos Q.$$

$$x^2/y^2 = r^2 + s^2 - 2rs \cos Q.$$

Therefore $(r - s)^2 < x^2/y^2 < (r + s)^2$. If at least one of r, s < 1/2, then $x^2/y^2 < 1$ and x < y, contradicting the fact that x $\geq$ y. If both r and s are equal to 1/2, then $x^2/y^2 = 1/2 - 1/2 \cos Q$, which is less than 1 since 1 $-$ cos Q < 2. Then again we get x < y, contradicting the fact that x $\geq$ y.

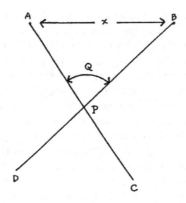

Fig. A1.

Appendix B

Proposition 3. The opposed parastichy pair (m, n) is visible if and only if the base of its opposed parastichy triangle has length one.

Proof.

1. Let (m, n) be a visible opposed parastichy pair. Assume that its opposed parastichy triangle has base $i > 1$ and vertex Q (see Fig. B1).

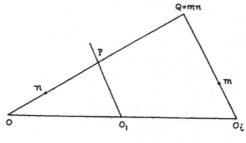

Fig. B1.

The leaf at Q is mn. Through 0_1 draw the parastichy that is parallel to the right leg of the triangle. It will intersect the left leg at some point P between 0 and Q. Since (m, n) is a visible opposed parastichy pair, there is a leaf at P. Call its leaf number x. Since P is on 0_1P which represents the same parastichy as 0_iQ, x is a multiple of m. Since P is on $0Q$, x is a multiple of n. Since m and n are relatively prime, x is divisible by mn, and therefore $x \geq mn$. This is impossible, since P is between 0 and Q and therefore must have a leaf number that is less than mn. Therefore the base of the triangle has length one.

2. Conversely, let the opposed parastichy triangle belonging to (m,n) have base 1 (Fig. B2).

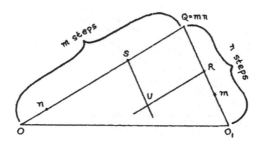

Fig. B2.

By the definition of opposed parastichy triangle, leaves n, 2n, 3n, ..., mn are on 0Q, and leaves m, 2m, 3m, ..., nm are on 0_1Q. The least residues modulo m of the numbers n, 2n, 3n, ..., mn are the numbers 0, 1, ..., m−1 rearranged, and the least residues modulo n of the numbers m, 2m, 3m, ..., nm are the numbers 0, 1, ..., n−1 rearranged. Consequently, in the opposed parastichy pair (m, n), each of the right parastichies passes through one of the leaves n, 2n, ..., mn on 0Q, and each of the left parastichies passes through one of the leaves m, 2m, ..., mn on 0_1Q. The right parastichy through mn and the left parastichy through mn have a leaf at their intersection. Choose any other right parastichy and any other left parastichy intersecting at some point U. The right parastichy passes through a leaf on 0_1Q at some point R. The left parastichy passes through a leaf on 0Q at some point S. QRSU is a parallelogram with leaves at Q, R and S. Therefore there is a leaf at U.

Each parastichy has many copies in the plane. More generally, any copy of a left parastichy must intersect the line 0Q at a leaf S representing a multiple of n (not necessarily on the segment 0Q), and any copy of a right parastichy must intersect the line 0_1Q at a leaf R representing a multiple of m (not necessarily on the segment 0_1Q). Let U be the intersection of these copies of the parastichies. Then, since in parallelogram QSRU Q, S and R are leaves, then so is U (This takes care of the fact that as a left and right parastichy spiral around the cylinder they intersect more than once).

Appendix C

Proposition 4. The contraction of a visible opposed parastichy pair is a visible opposed parastichy pair.

Proof. Assume (m, n) with m > n is a visible opposed parastichy pair. The visible opposed parastichy triangle belonging to (m, n) (Fig. C1) has the following properties: Leaf n is on the left leg, and there is no leaf between 0 and n. There are m steps on the left leg. Leaf m is on the right leg, and there is no leaf between 0_1 and m. There are n steps on the right leg. Leaf mn is at the vertex P. The base has length 1. On P0 count down n steps from P to leaf T whose number is (m − n)n. Draw the

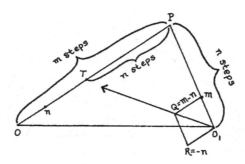

Fig. C1.

parastichy through 0_1 that is determined by n. Extend it away from n. Let R be the point where leaf − n is located. Let Q be the fourth vertex of the parallelogram that has vertices at 0_1, R and m. There is a leaf at Q whose number is m − n. Moreover, there is no leaf between 0_1 and Q, since there is no leaf in the interior of the parallelogram. Leaf (m - n)n lies on the line 0_1Q, say at point T'. There are n steps between 0_1 and T'. We now prove that T and T' coincide. Let a be the length of the projection of 0n on 00_1. Let b be the length of the projection of 0_1m on 00_1. ma + nb = 1. The length of the projection of 0_1Q on 00_1 is a + b. The length of the projection of 0T on 00_1 is (m − n)a. The length of the projection of 0_1T' on 00_1 is n(a + b). T and T' are at the same height above 00_1, since they represent the same leaf n(m − n). They coincide if and only if the sum of the lengths of the projections of 0T and 0_1T' is 1. But (m − n)a + n(a + b) = ma + nb = 1. Therefore $0T0_1$ is a visible opposed parastichy triangle belonging to the visible opposed parastichy pair (m − n, n).

Proposition 6. For every t > 1, (t, t + 1) is a visible opposed parastichy pair if and only if $1/(t + 1) \le d \le 1/t$.

Proof.

1. Assume $1/(t + 1) < d < 1/t$ (Fig. C2). Since td < 1, and 1 − td < d < 1/t, leaf t is to the left of 0_1. Since (t + 1)d > 1, and (t + 1)d − 1 < d < 1/t, leaf t + 1 is to the right of 0. Draw the parastichy through 0 and t + 1 for t steps to Q. Draw the parastichy through 0_1 and t for t + 1 steps to Q'. We show that Q and Q' coincide. The projection of 0Q on 00_1 has length t[(t + 1)d − 1]. The projection of 0_1Q' on 00_1 has length (t + 1)(1 − td). Their sum has length t[(t + 1)d − 1] + (t + 1)(1 − td) = 1. Therefore Q and Q' coincide, and hence (t, t + 1) is a visible opposed parastichy pair. In the limiting cases where d = 1/t or d = 1/(t + 1) the argument still holds.

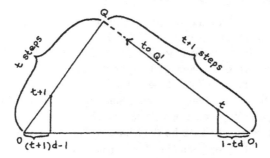

Fig. C2.

2. Conversely, if (t, t + 1) is a visible opposed parastichy pair, (t + 1)d − 1 ≥ 0 implies that $d \ge 1/(t + 1)$, and 1 − td ≥ 0 implies that $d \ge 1/t$.

Let (v, y) be a visible opposed parastichy pair, with associated visible opposed parastichy triangle as shown in Fig. C3. Before we undertake to prove Proposition 7, we first observe some properties of the triangle, and prove a useful lemma.

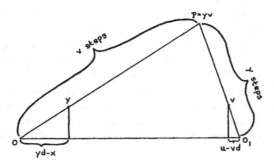

Fig. C3.

Let $x = (yd)$ = nearest integer to yd. We have $x \leq yd$. Let $u = (vd)$ = nearest integer to vd. We have $u \geq vd$. The projection of $0y$ on 00_1 has length $yd - x \geq 0$. The projection of 0_1v on 00_1 has length $u - vd \geq 0$. Therefore $x/y \leq d \leq u/v$. That is, the maximum range of possible values of d is the interval $[x/y, u/v]$. The projection of $0P$ on 00_1 has length $v(yd - x)$. The projection of 0_1P on 00_1 has length $y(u - vd)$. Therefore $v(yd - x) + y(u - vd) = 1$. Consequently $yu - vx = 1$. This implies that x and y are relatively prime, and u and v are relatively prime. Then the mediant m between x/y and u/v is $(x + u)/(y + v)$.

Lemma. If the left extension or the right extension of a visible opposed parastichy pair is opposed, (that is, consists of a set of left parastichies and a set of right parastichies), then it is a visible opposed parastichy pair.

Proof (for a left extension). Let (v, y) be a visible opposed parastichy pair. Draw its visible opposed parastichy triangle $0P0_1$. Extend $0P$ to Q by adding y more steps as in Fig. C4. The leaf at Q is $(v + y)y$. Through 0_1 draw the parastichy to leaf y. The fourth vertex S of the parallelogram that has vertices at 0_1, v and y is $v + y$. Then there is a point on 0_1S whose leaf number is $y(v + y)$. Call this point Q'. There is no leaf between between 0_1 and S since there is no leaf in the interior of parallelogram 0_1vSy. We now show that Q' and Q coincide. Let a be the length of the projection of $0y$ on 00_1; let b be the length of the projection of 0_1v on 00_1; we have $va + yb = 1$. The projection of 0_1S on 00_1 has length $b - a$. The projection of $0_1Q'$ on 00_1 has length $y(b - a)$. The projection of $0Q$ on 00_1 has length $(v + y)a$. $(v + y)a + y(b - a) = va + yb = 1$. Therefore Q' and Q coincide, and triangle $0Q0_1$ is a visible opposed parastichy triangle belonging to the visible opposed parastichy pair $(v + y, y)$. A similar proof applies to the right extension $(v, v + y)$.

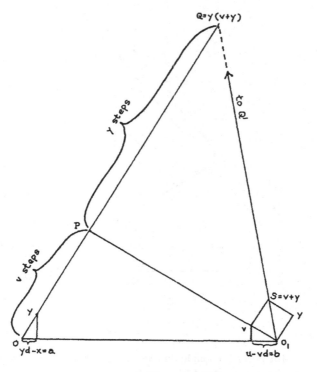

Fig. C4.

Proposition 7. Suppose, as in Fig. C3, that $[x/y, u/v]$ is the range of all possible values of d for which the opposed parastichy pair (v, y) is visible. x/y and u/v are in lowest terms. Let m be the mediant between x/y and u/v, namely $m = (x + u)/(y + v)$. Then the left extension of (v, y) is a visible opposed parastichy pair if and only if d is in the segment $[x/y, m]$, and the right extension of (v, y) is a visible opposed parastichy pair if and only if d is in the segment $[m, u/v]$.

Proof (for a left extension). See Fig. C5 where we extended OP by y steps to Q. Assume that the extension $(v + y, y)$ is an opposed parastichy pair, and hence, by the Lemma, a visible opposed parastichy pair. We show that $x/y \leq d \leq m$. The projection of $0_1 S$ on 00_1 has length $(u - vd) - (yd - x)$. $u - vd - yd + x \geq 0$. Therefore $d \leq (u + x)/(v + y) = m$, $yd - x \geq 0$, and therefore $d \geq x/y$. Hence d is in the segment $[x/y, m]$. Conversely, if $d \leq (u + x)/(v + y)$, then $(v + y)d \leq u + x$, and $0 \leq (u - vd) - (yd - x)$, which implies that S is to the left of the vertical line through 0_1, and hence $(v + y, y)$ is opposed. A similar proof can be carried out for the right extension of (v, y).

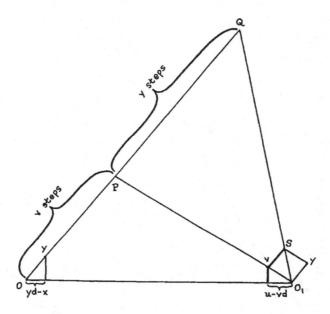

Fig. C5.

Appendix D

Proposition 8. If r is small enough, and the minimum distance between leaves is maximized, then the leaves nearest 0 are equidistant from it.

Proof. Let v and y be the two leaves nearest 0. Because (v, y) is a conspicuous opposed parastichy pair, it is a visible opposed parastichy pair. (Proposition 2). The visible opposed parastichy triangle belonging to it is as in Fig. C3. We have

$$\text{dist}^2(0, y) = (yd - x)^2 + y^2 r^2.$$

$$\text{dist}^2(0, v) = (u - vd)^2 + v^2 r^2.$$

If for fixed r we plot dist^2 as a function of d, each of these equations yields a parabola with vertical axis like those shown in Fig. 6. If r is small enough, the two parabolas intersect at a point where the value of d has meaning in a phyllotaxis context (namely, $0 < d \leq 1/2$). It is clear from the graph that $\text{dist}^2(0, y)$ and $\text{dist}^2(0, v)$ are both maximized at the point where the parabolas intersect. If either, say $\text{dist}^2(0, y)$, is made larger by a further change in d, the corresponding point on the parabola is above the parabola for $\text{dist}^2(0, v)$, and $\text{dist}(0, y)$ ceases to be the minimum distance between leaves.

When is r "small enough?" There is a separate answer for each region $1/(t + 1) \leq d \leq 1/t$. We have already answered the question for $t = 2$ in Sec. 14, namely that r must be less than or equal to the square root of $1/12$. We consider now t > 2. In this region, the first visible opposed parastichy pair capable of becoming conspicuous is $(t, 1)$. We determine the least upper bound for r that permits $\text{dist}^2(0, 1) = \text{dist}^2(0, t)$. We have $\text{dist}^2(0,1) = d^2 + r^2$, and $\text{dist}^2(0, t) = (1 - td)^2 + t^2r^2$. Equating them yields the equation $(d - t/(t^2-1))^2 + r^2 = (1/(t^2 - 1))^2$, $r > 0$. This defines a semicircle in the (d, r) plane whose center is at $(t/(t^2- 1), 0)$ and whose radius is $1/(t^2- 1)$. In this case "r is small enough" means $r \leq 1/(t^2- 1)$.

As r decreases further while the minimum distance between leaves is maximized, the point (d, r) descends along this semicircle until a third leaf is near enough to 0 to displace either 1 or t as the leaf nearest 0. Then the point (d, r) will descend along a different semicircle. We now answer two questions: Which leaf will that third leaf be? What will be the new path of (d, r) be? We answer these questions more generally for descent along any (v, y) semicircle.

The Addition Rule. Under the condition that the minimum distance between leaves is maximized, and r is decreasing, the transition from (v, y) phyllotaxis with $v < y$ to the next higher phyllotaxis must be to $(v+y, y)$.

Proof. Let M be the leaf with leaf number v. Let N be the leaf with leaf number y. Assume that there is a leaf P with leaf number p that has just joined M and N as leaves nearest 0 (Fig. D1). Then we have $0M = 0N = 0P$. Let 1 be the element of the cylinder through leaf 0. Point P is closer to 1 than M or N. Points M and N are

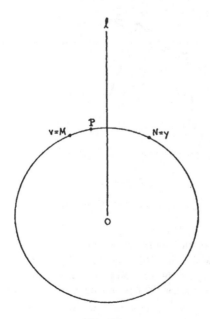

Fig. D1.

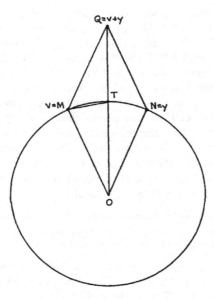

Fig. D2.

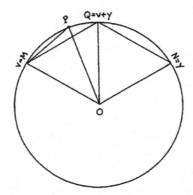

Fig. D3.

on opposite sides of l in the plane development of the cylinder. Point P is on the arc MN of the circle with center 0 and radius 0M. Complete the parallelogram on 0MN. Let the fourth vertex be Q. Point Q is a leaf with leaf number v + y. We prove that P = Q. We consider the possible positions of Q with respect to the circle.

1) Q is not inside the circle, since then we would have 0Q < 0M, contradicting the fact that 0M is the minimum distance between leaves.

2) If Q is outside the circle, let T be the point of the circle that is on 0Q (Fig. D2). Angle MT0 > angle MQ0. Since MQ = N0 = M0, angle MQ0 = angle M0Q. Therefore angle MT0 > angle M0Q, and 0M > MT. If P = T, we have 0M > MP, contradicting the fact that 0M is the minimum distance between leaves. If P is not T, we may assume without loss of generality that P is between M and T. Then we have MT > MP, and 0M > MP, again a contradiction.

3) If Q is on the circle and P is not Q, assume without loss of generality that P is between M and Q on the arc MQN (Fig. D3).

Angle M0P < angle M0Q. Therefore MP < MQ, again contradicting the fact that MQ (which equals M0) is the minimum distance between leaves. Therefore we have P = Q, and $p = v + y$.

In the proof above we assumed that there is a leaf capable of joining v and y as leaves nearest 0, and we showed that, if so, it must be $v + y$. This is so only if the semicircle defined by $\text{dist}^2(0, v) = \text{dist}^2(0, y)$, $r > 0$, intersects the semicircle defined by $\text{dist}^2(0, v + y) = \text{dist}^2(0, y)$, $r > 0$. (If those two intersect, they also intersect the semicircle defined by $\text{dist}^2(0, v + y) = \text{dist}^2(0, v)$, $r > 0$). We now show that they do indeed intersect. Taking data from Fig. C3, we obtain as the equation for the (v, y) semicircle,

$$yd - (xy - uv)/(y^2 - v^2))^2 + r^2 = 1/(y^2 - v^2)^2, r > 0,$$

taking into account that $yu - vx = 1$. The center of the semicircle is at $((xy - uv)/(y^2 - v^2), 0)$ and the radius is $1/(y^2 - v^2)$. The upper and lower intercepts of this semicircle on the d axis are respectively equal to $(x + u)/(y + v)$ and $(x - u)/(y - v)$, taking into account that $uy - vx = 1$.

The equation for the (v + y, y) semicircle is

$$(d - (uv + uy + xv)/(v^2 + 2vy))^2 + r^2 = 1/(v^2 + 2vy)^2, r > 0.$$

The center is at $((uv + uy + xv)/(v^2 + 2vy), 0)$, and the radius is $1/(v^2 + 2vy)$. The upper and lower intercepts on the d axis are respectively u/v and $(2x + u)/(2y + v)$. Taking into account that $uy - vx = 1$, we find that the intercepts of the two semicircles are arranged in the following order:

$$(x - y)/(y - v) < (2x + u)/(2y + v) < (x + u)/(y + v) < u/v.$$

These inequalities imply that the two semicircles intersect, since each joins a point in the interior of the other circle to a point outside it.

When the point (d, r), descending on the (v, y) semicircle, reaches the point where it intersects both the (v + y, y) semicircle and the (v, v + y) semicircle, which one will it switch to? We know that for (v, y) to be a visible opposed parastichy pair, we have $x/y \leq d \leq u/v$. For (v + y, y) to be a visible opposed parastichy pair, we have $x/y \leq d \leq (u + x)/(v + y)$. For (v, v + y) to be a visible opposed parastichy pair, we have $(u + x)/(v + y) \leq d \leq u/v$. But while the point (d, r) is on the (v, y) semicircle, d must be between its intercepts on the d axis. That is,

$$(x - u)/(y - v) < d < (x + u)/(y + v).$$

This condition is inconsistent with the condition for the (v, v + y) semicircle. Therefore (d, r) must switch to the (v + y, y) semicircle.

Appendix E. Computer Program for the Cylindrical Representation of a Disc Model

T = time measured in plastochrones; X and Y are leaves nearest 0, so that (X, Y) or (Y, X) is the phyllotaxis; D is the divergence.

```
100 LPRINT "T", "X", "Y", "D":CLS
150 LINE (0,0)–(0,195)
170 LINE (0,195)–(595,195)
200 INPUT "A=";A:INPUT "D=";D
210 T=2
215 S=10:X=0:Y=0
220 R=[LOG((1+A*T)/(1+A*(T-1)))]/2*3.1415926)
230 For N=1 to T:V=N:GOSUB 400
240 IF W<S THEN 380
250 NEXT N
300 I=.1*k:FOR J=1 TO 4
310 I=.1*I
315 D=D+I
320 V=Y:GOSUB 400:M=W
330 V=X:GOSUB 400:L=W
340 IF M−L>=0 THEN 315
350 D=D−I:NEXT J
360 GOTO 500
370 T=T+1:GOTO 215
380 S=W:Y=X:X=N:K=E:GOTO 250
400 F=V*D−INT(V*D)
410 IF F<=.5 THEN 570
420 IF F>.5 THEN 580
430 W=V*V*R*R+H*H
440 RETURN
500 D=D+.5*I
510 V=Y:GOSUB 400:M=W
520 V=X:GOSUB 400:L=W
530 IF M−L<0 THEN 542
540 IF M−L>=0 THEN 545
542 D=D−.5*I:GOTO 550
545 D=D+.5*I:GOTO 550
550 PSET (1500*D,195−10000*R):LPRINT T,X,Y,D
560 GOTO 370
570 H=F:E=1:GOTO 430
580 H=1−F:E=−1:GOTO 430
```

Table E1. r=[ln $\{(1 + aT) / [1 + a(T - 1)]\}$] /(2 * 3.1415926),
a =0.14, T_c =2, initial d is between 1/3 and 1/2; nr is used as
vertical component of dist(0, n). (Cylindrical of disc model.)

T	d	Phyllotaxis	T	d	Phyllotaxis
2	0.33384	(1, 2)	35	0.38175	(8,13)
3	0.39932	(3, 2)	36	0.38171	(8,13)
4	0.39944	(3, 2)	37	0.38168	(8,13)
5	0.37652	(3, 5)	38	0.38164	(8,13)
6	0.37628	(3, 5)	39	0.38161	(8,13)
7	0.37610	(3, 5)	41	0.38156	(8,13)
8	0.38220	(8, 5)	42	0.38153	(8,13)
9	0.38251	(8, 5)	43	0.38151	(8,13)
10	0.38276	(8, 5)	44	0.38149	(8,13)
11	0.38298	(8, 5)	45	0.38146	(8,13)
12	0.38315	(8, 5)	46	0.38145	(8,13)
13	0.38330	(8, 5)	47	0.38143	(8,13)
14	0.38343	(8, 5)	48	0.38141	(8,13)
15	0.38354	(8, 5)	49	0.38139	(8,13)
16	0.38363	(8, 5)	50	0.38138	(8,13)
17	0.38372	(8, 5)	56	0.38133	(21,13)
18	0.38348	(8, 13)	60	0.38147	(21,13)
19	0.38325	(8, 13)	70	0.38171	(21,13)
20	0.38306	(8, 13)	80	0.38186	(21,13)
21	0.38289	(8, 13)	90	0.38196	(21,13)
22	0.38274	(8, 13)	100	0.38204	(21,13)
23	0.38261	(8, 13)	110	0.38209	(21,13)
24	0.38249	(8, 13)	120	0.38213	(21,13)
25	0.38239	(8, 13)	130	0.38216	(21,13)
26	0.38230	(8, 13)	140	0.38219	(21,13)
27	0.38221	(8, 13)	150	0.38221	(21,13)
28	0.38213	(8, 13)	156	0.38222	(21,34)
29	0.38206	(8, 13)	160	0.38220	(21,34)
30	0.38200	(8, 13)	170	0.38215	(21,34)
31	0.38194	(8, 13)	180	0.38211	(21,34)
32	0.38189	(8, 13)	190	0.38208	(21,34)
33	0.38184	(8, 13)	200	0.38205	(21,34)
34	0.38179	(8, 13)		d converges to τ^{-2}	

Table E2. $r = (4 \times 3.1415926)^{-1}[1 + 4a(T + b)]^{1/2}\ln([T + b]/[T + b - 1])$, $a = 0.1$, $b = 7$; nr is used as vertical component of dist$(0, n)$. (Cylindrical representation of parabolic model); $T_c = 5$; initial d is between 2/5 and 1/2.

T	d	Phyllotaxis	T	d	Phyllotaxis
5	0.4256	(5, 2)	19	0.4180	(5, 7)
6	0.4259	(5, 2)	20	0.4180	(5, 7)
			30	0.4176	(5, 7)
7	0.4195	(5, 7)			
8	0.4193	(5, 7)	40	0.4178	(12, 7)
9	0.4191	(5, 7)	50	0.4185	(12, 7)
10	0.4189	(5, 7)	60	0.4190	(12, 7)
11	0.4188	(5, 7)	70	0.4193	(12, 7)
12	0.4186	(5, 7)	80	0.4195	(12, 7)
13	0.4185	(5, 7)	90	0.4197	(12, 7)
14	0.4184	(5, 7)	100	0.4198	(12, 7)
15	0.4183	(5, 7)	200	0.4204	(12, 7)
16	0.4182	(5, 7)			
17	0.4182	(5, 7)	300	0.4204	(12, 19)
18	0.4181	(5, 7)	400	0.4201	(12, 19)

d converges to
$$\frac{1}{2 + \dfrac{1}{2 + \tau^{-1}}}$$

References

Adler I., A model of contact pressure in phyllotaxis, *J. Theor. Bio.* **45** (1974) 1–79.

Adler I., The consequences of contact pressure in phyllotaxis, *J. Theor. Biol.* **65** (1974) 29–77.

Airy H., On leaf arrangement, *Proc. R. Soc.* **1**, **21**, (1873) 176–179.

Bravais L. & A., Essai sur la disposition des feuilles curviseriees, *Annals. Sci. Nat. Botanique* **(2) 7** (1837) 42–110: 8, 11.

Jean R. V., *Mathematical Approach to Pattern & Form in Plant Growth* (Wiley-Interscience, New York, 1984).

Schwendener S. *Mechanische Theorie der Blattstellungen* (Engelmann, Leipzig, 1878).

Irving Adler: Vita and Afterword

- Born in New York City, April 27, 1913
  Married Ruth Relis, 1935 (deceased 1968); children Stephen and Peggy
  Married Joyce Theresa Lifshutz, 1968 (deceased 1999)

- Primary and secondary education: New York City public schools
  B.S., City College of New York, 1931, magna cum laude
  M.A. in mathematics, Columbia University, 1938
  Ph.D. in mathematics, Columbia University, 1961

- Teacher of mathematics, New York Public Schools, 1932–1952
  Chairman of mathematics department, Textile High School, 1946–1952

- Author of 56 books on mathematics, science, and education; coauthor
  (with Ruth Adler) of 30 science books for younger children

- Fellow, American Association for the Advancement of Science, 1982
  Townsend Harris Medal, City College Alumni Association, 1993

- Honorary Doctorates:
  St. Michael's College, Winooski, Vermont (D. Sc.), 1990
  City College of the City University of New York (D. H. L), 2002

- American Civil Liberties Union of Vermont Lifetime Achievement
  Award, 2009

- Author of research papers on phyllotaxis, reprinted in this volume.

- Irving Adler's afterword: "In the course of developing my model,
  I discovered and proved what came to be known as the fundamental
  theorem of phyllotaxis. Since my model begins with explicitly stated
  assumptions and draws conclusions from them rigorously, it is a piece
  of pure mathematics. As such, it may be thought of as an addition to
  Hermann Minkowski's geometry of numbers."